Mathematical Muffin Morsels
Nobody wants a small piece

Problem Solving in Mathematics and Beyond

Print ISSN: 2591-7234
Online ISSN: 2591-7242

Series Editor: Dr. Alfred S. Posamentier
Distinguished Lecturer
New York City College of Technology - City University of New York

---

There are countless applications that would be considered problem solving in mathematics and beyond. One could even argue that most of mathematics in one way or another involves solving problems. However, this series is intended to be of interest to the general audience with the sole purpose of demonstrating the power and beauty of mathematics through clever problem-solving experiences.

Each of the books will be aimed at the general audience, which implies that the writing level will be such that it will not engulfed in technical language — rather the language will be simple everyday language so that the focus can remain on the content and not be distracted by unnecessarily sophiscated language. Again, the primary purpose of this series is to approach the topic of mathematics problem-solving in a most appealing and attractive way in order to win more of the general public to appreciate his most important subject rather than to fear it. At the same time we expect that professionals in the scientific community will also find these books attractive, as they will provide many entertaining surprises for the unsuspecting reader.

Published

Vol. 16 *Mathematical Muffin Morsels: Wants a Small Piece*
by William Gasarch, Erik Metz, Jacob Prinz and Daniel Smolyak

Vol. 15 *Teaching Secondary School Mathematics: Techniques and Enrichment*
by Alfred S Posamentier and Beverly Smith

Vol. 14 *Engaging Young Students in Mathematics through Competitions — Perspectives and Practices: Volume II — Mathematics Competitions and how they relate to Research, Teaching and Motivation; Entertaining and Informative Papers from the WFNMC8 Congress in Semriach/Austria 2018*
edited by Robert Geretschläger

Vol. 13 *Engaging Young Students in Mathematics through Competitions — World Perspectives and Practices: Volume I — Competition-ready Mathematics; Entertaining and Informative Problems from the WFNMC8 Congress in Semriach/Austria 2018*
edited by Robert Geretschläger

For the complete list of volumes in this series, please visit www.worldscientific.com/series/psmb

Problem Solving in Mathematics and Beyond — Volume 16

Mathematical Muffin Morsels
Nobody wants a small piece

William Gasarch
Erik Metz
Jacob Prinz
Daniel Smolyak
University of Maryland, USA

World Scientific

NEW JERSEY · LONDON · SINGAPORE · BEIJING · SHANGHAI · HONG KONG · TAIPEI · CHENNAI · TOKYO

Published by

World Scientific Publishing Co. Pte. Ltd.
5 Toh Tuck Link, Singapore 596224
USA office: 27 Warren Street, Suite 401-402, Hackensack, NJ 07601
UK office: 57 Shelton Street, Covent Garden, London WC2H 9HE

Library of Congress Cataloging-in-Publication Data
Names: Gasarch, William I., author. | Metz, Erik, author. | Prinz, Jacob, author. | Smolyak, Daniel, author.
Title: Mathematical muffin morsels : nobody wants a small piece / William Gasarch (University of Maryland, USA), Erik Metz (University of Maryland, USA), Jacob Prinz (University of Maryland, USA) and Daniel Smolyak (University of Maryland, USA).
Description: New Jersey : World Scientific, [2020] | Series: Problem solving in mathematics and beyond, 2591-7234 ; vol. 16 | Includes bibliographical references and index.
Identifiers: LCCN 2020007589 | ISBN 9789811215179 (hardcover) | ISBN 9789811215971 (paperback) | ISBN 9789811215186 (ebook) | ISBN 9789811215193 (ebook other)
Subjects: LCSH: Mathematics--Problems, exercises, etc. | Division--Problems, exercises, etc. | Fractions--Problems, exercises, etc.
Classification: LCC QA43 .G347 2020 | DDC 513.2/1--dc23
LC record available at https://lccn.loc.gov/2020007589

British Library Cataloguing-in-Publication Data
A catalogue record for this book is available from the British Library.

Copyright © 2020 by World Scientific Publishing Co. Pte. Ltd.

All rights reserved. This book, or parts thereof, may not be reproduced in any form or by any means, electronic or mechanical, including photocopying, recording or any information storage and retrieval system now known or to be invented, without written permission from the publisher.

For photocopying of material in this volume, please pay a copying fee through the Copyright Clearance Center, Inc., 222 Rosewood Drive, Danvers, MA 01923, USA. In this case permission to photocopy is not required from the publisher.

For any available supplementary material, please visit
https://www.worldscientific.com/worldscibooks/10.1142/11689#t=suppl

Desk Editors: V. Vishnu Mohan/Tan Rok Ting

Typeset by Stallion Press
Email: enquiries@stallionpress.com

Preface

1. How I Found the Muffin Problem

While writing my earlier book *Problems with a Point: Exploring Math and Computer Science* (with co-author Clyde Kruskal), my contact at World Scientific, Rochelle Kronzek, got me an invitation to *The Twelfth Gathering for Gardner* (2016). This was a meeting to celebrate the life and works of Martin Gardner, who wrote a column on mathematical recreations in *Scientific American* for many years. I saw many talks of interest; however, the most interesting thing I saw was a problem in a booklet:

> **Julia Robinson Mathematics Festival:**
> **A Sample of Mathematical Puzzles**
> *Compiled by Nancy Bachman*

Begin Excerpt

> **The Muffin Puzzle**
> Invented by Recreational Mathematician Alan Frank
> Described by Jeremy Copeland in
> The New York Times Numberplay Online Blog
> wordplay.blogs.nytimes.com/2013/08/19/cake

You have 5 muffins and 3 students. You want to divide the muffins evenly, but no student wants a tiny sliver. What division of muffins maximizes the smallest piece?

Here is a picture of the muffins and a possible division:

I call the students RED, BLUE, and GREEN.

- RED gets the first muffin and $\frac{2}{3}$ of the fourth muffin.
- BLUE gets the second muffin and $\frac{2}{3}$ of the fifth muffin.
- GREEN gets the third muffin, $\frac{1}{3}$ of the fourth muffin, and $\frac{1}{3}$ of the fifth muffin.

GREEN gets a piece of size $\frac{1}{3}$, which she thinks is small since she has big hands. Is there a division with a larger smallest piece?

Here are some other questions to consider:

- How would you divide 3 muffins between 5 students?
- How would you divide 4 muffins between 7 students?

End Excerpt

I solved those three problems. That was fun! I then changed the problem from (the so-called) *recreational math* to (the so-called) *serious math* by replacing the numbers with variables:

You have m muffins and s students. You want to divide the muffins evenly, but no student wants a tiny sliver. What division of muffins maximizes the smallest piece?

2. Let's Write a Book!

Over the next three years, I investigated this problem. I had help from a lot of people (mostly students). The following sequence of events summarizes the process:

(1) used our techniques to solve several problems;
(2) found a problem that our techniques do not solve;
(3) found new techniques to solve that problem, often with help;
(4) added these new techniques and a co-author;
(5) Lather. Rinse. Repeat.

Out of this work came a paper [Cui *et al.* (2018)] and a survey [Gasarch *et al.* (2019)]. This work covers a small fraction of what we have done. So, a subset of my fellow researchers (Erik Metz, Jacob Prinz, Daniel Smolyak) decided to gather up *all* of our works (and some other works) on the muffin problem into a book. As we wrote the book, we realized that to keep it simple and coherent, some material would have to be omitted. We decided to make a website for this extra material. *The MUFFIN website* is:

www.cs.umd.edu/users/gasarch/MUFFINS/muffins.html.

There have also been some excellent works by other people, which we had to omit. In particular, Scott Huddleston had a wonderful method that seemed to solve *every muffin problem quickly*, though he had no proof of this. Richard Chatwin (2019) independently discovered Scott's method and proved that it always works. We describe Scott's algorithm in Chapter 13; however, we omit Richard's proof that it works.

3. Our Intended Audience

A large proportion of this book (I would guess $\frac{121}{136}$) requires no mathematics beyond elementary algebra. Hence, I am tempted to say that *any high school student with an interest in mathematics could read most of this book.* While I believe that is true, here are some additional thoughts:

(1) A high school student *without* an interest in math could read most of this book and *develop* an interest in math. I've heard stories of parents who give their children puzzle books that are really math, but so long as they don't call it math, the children enjoy it. This book may be in that category. Let's hope no such children read this preface.
(2) The reader may not know some of the mathematical notation we use; however, we summarize it in Appendix A.
(3) It takes some mathematical maturity to read this book. There is an alternative viewpoint: reading this book will give one mathematical maturity.

There are two chapters and one appendix that require a bit more mathematical sophistication. They are starred (*).

4. Fair Division: A Related Field

When I talk about this topic, I am sometimes asked how it relates to the field of *Fair Division*. I would call it a close cousin of that field. Appendix B is a summary of some of the works in Fair Division.

5. Prior Work on the Muffin Problem

While working on the problem, I searched on Google to see if anyone else had worked on it. I never found anything. However, in 2017, James Propp informed me that Alan Frank had originally posed The Muffin Problem within a *private* math email group in 2009. You have to be a member of the group to see their discussions, which is why my Google searches turned up empty.

James emailed me all of the emails the group generated about the problem. They mostly went in a different direction from our work; however,

(1) Theorem 2.9 was proved by Erich Friedman.
(2) Theorem C.9 was proved by Veit Elser.
(3) Theorem C.14 was proved by Caleb Stanford.
(4) They conjectured that if the number of muffins is more than the number of students, then there is a fair allocation with smallest piece $\geq \frac{1}{3}$.

The first two theorems were proven by me and my co-authors and I, though 7 years after Erich Friedman and Veit Elser. The third theorem was new to me and my co-authors. The conjecture was proven by Erik Metz and is on the above-mentioned website. Later, Richard Chatwin (2019) came up with an alternative proof.

6. A Celebration

James Propp, Alan Frank, and I agreed to all meet in Boston and arranged it for when I would give a talk on The Muffin Problem to the MIT combinatorics seminar. Alan Frank brought five muffins: one cut $\{\frac{1}{2}, \frac{1}{2}\}$ and the rest cut $\{\frac{5}{12}, \frac{7}{12}\}$. Alan, James, and I used these pieces to give each of us $\frac{5}{3}$.

William Gasarch

About the Authors

William Gasarch (aka Bill) received his B.S. in Math and Applied Math (well-rounded) from SUNY Stony brook in 1980 and a Ph.D. in Computer Science from Harvard in 1985. He has been a professor at the University of Maryland, in Computer Science, from 1985 until the present. He has worked in Computational Complexity, Computability Theory, Computational Learning Theory, Ramsey Theory, and of course Muffins. He has published over 50 papers. He has mentored over 100 high school students and over 50 undergraduates on research projects. Bill has been co-blogging with Lance Fortnow on http://blog.computationalcomplexity.org since 2007. He had previously written two books: (1) *Bounded Queries in Recursion Theory* (co-authored with Georgia Martin) and (2) *Problems with a Point* (co-authored with Clyde Kruskal). He has a muffin paper in the Conference on Fun with Algorithms. He collects novelty songs and has the largest collection of satires of Nobel Laureate Bob Dylan.

Erik Metz received his B.S. in Mathematics and Computer Science from the University of Maryland at College Park in the spring of 2020. He will be working in Finance starting in the fall of 2020. He has conducted research on Extremal Combinatorics, Financial Mathematics, and of course Muffins. He has published in *the European Journal of Combinatorics* and has a muffin paper in the Conference on Fun with Algorithms. In the past, he had interned at the Johns Hopkins Applied Physics Lab and attended the University of Minnesota Duluth summer program for undergraduate research in mathematics.

Jacob Prinz received his B.S. in Mathematics from the University of Maryland at College Park in the spring of 2020. He entered the University of Maryland Graduate School in Computer Science in the fall of 2020 and plans to work on type theory. He has conducted research in Probability, Type Theory, and of course Muffins. He has a muffin paper in the Conference on Fun with Algorithms. In the past, he had attended the University of Indiana summer program for undergraduate research in mathematics.

Daniel Smolyak received his B.S. in Economics and Computer Science from the University of Maryland at College Park in the spring of 2020. He entered the University of Maryland Graduate School in Computer Science in the fall of 2020 and plans to work on economics and computation. He has conducted research in the fields of Human-Computer Interaction, Data Science,

Natural Language Processing, and of course muffins. He has worked as a software engineering intern at the Johns Hopkins Applied Physics Laboratory and Microsoft Corporation. He has published works in the ACM CHI Conference, the IEEE Big Data Conference, and has a muffin paper in the Conference on Fun with Algorithms.

Acknowledgments

We thank Rochelle Kronzek, since it was her getting Bill an invite to the Gardner conference that got the research going.

We thank Alan Frank for coming up with the problem and for making a conjecture that seems to be true (see Section 4.9). We thank Jim Propp for providing us with information on the prior works on the muffin problem. We also thank Veit Elser, Erich Friedman, and Caleb Stanford for their prior works in the area. Caleb helped us write the section on his work.

We would like to thank Stephanie Warman for her awesome muffin code. Much of the code at the MUFFIN website is hers.

We thank Guangqi Cui, John Dickerson, Naveen Durvasula, Naveen Raman, and Sung Hynn Yoo who helped us develop some of the techniques. They are co-authors on a muffin paper [Cui *et al.* (2018)] that appeared in the *Fun with Algorithms* conference. We thank Saadiq Shaik and Daniel Wei who wrote some code for us. We thank Ken Tan who worked out, by hand, some formulas for us. We thank Scott Huddleston and Richard Chatwin for sharing their excellent work on the muffin problem with us.

We thank Sofia Bzhilyanskaya, Doug Chen, Yunseo Choi, Nathan Grammel, Lexa Hummel, Clyde Kruskal, Patrick O'Toole, Sachin Pandey, and Alex Zhang for the helpful discussions. We thank Doug Chen, Yunseo Choi, Nathan Grammel, Nathan Hayes, Lexa Hummel, David Marcus, Leenah Shalhoub, Patrick O'Toole, and Stephanie Warman for proofreading and checking the manuscript.

Bill thanks Lance Fortnow for helping write and perform the song *Muffin Math*, which you can see on YouTube:

https://www.youtube.com/watch?v=4xQFlsK7jKg.

Bill thanks Darling, who put up with him talking about muffins for 3 years. Co-authors Daniel, Erik, and Jacob thank their suite-mate Robby, who put up with them talking about muffins for 3 years.

Contents

Preface   v

About the Authors   ix

Acknowledgments   xiii

1. Five Muffins, Three Students; Three Muffins, Five Students   1
2. One Student! Two Students! Some Basic Theorems!   11
3. Our Plan   17
4. Three Students! Four Students! The Floor–Ceiling Theorem!   19
5. Finding Procedures   27
6. The Half Method   45
7. A Formula for $f(m, 5)$ 67
8. The Interval Method   75
9. The Midpoint Method   85
10. The Easy Buddy–Match Method   97

11.	The Hard Buddy–Match Method	115
12.	The Gap and Train Methods	139
13.	Scott Huddleston's Method	155

Appendix A: Math Notation	181
Appendix B: Fair Division	193
Appendix C: $f(m, s)$ Exists! $f(m, s)$ is Rational! $f(m, s)$ is Computable!	195
References	207
Index	209

Chapter 1

Five Muffins, Three Students; Three Muffins, Five Students

1.1. Five Muffins, Three Students

You have 5 muffins and 3 students. You want to divide the muffins evenly so that each student gets $\frac{5}{3}$ muffins. The following picture shows one way to do this:

- RED gets the first muffin and $\frac{2}{3}$ of the fourth muffin.
- BLUE gets the second muffin and $\frac{2}{3}$ of the fifth muffin.
- GREEN gets the third muffin, $\frac{1}{3}$ of the fourth muffin, and $\frac{1}{3}$ of the fifth muffin.

One of GREEN's pieces is of size $\frac{1}{3}$, which she thinks is small since she has big hands.

Exercise 1.1. Is there a way to divide 5 muffins into pieces, and give the pieces to 3 students, such that (1) every student gets $\frac{5}{3}$, and (2) every piece is *bigger* than $\frac{1}{3}$?

Try to solve this yourself before reading on.

Solution to Exercise 1.1
The following picture shows how to divide 5 muffins to give to 3 students so that (1) everyone gets $\frac{5}{3}$ and (2) the smallest piece is $\frac{5}{12}$:

- RED gets $\frac{1}{2}$ of the first muffin and $\frac{7}{12}$ of the second muffin and third muffin.
- BLUE gets $\frac{1}{2}$ of the first muffin and $\frac{7}{12}$ of the fourth muffin and fifth muffin.
- GREEN gets $\frac{5}{12}$ of the second, third, fourth, and fifth muffins.

End of Solution of Exercise 1.1

Is there a procedure where the smallest piece is $> \frac{5}{12}$? What do you think is true? What do you hope is true? We show that there is no procedure where the smallest piece is $> \frac{5}{12}$. We also include a formal description of the procedure with smallest piece $\frac{5}{12}$.

Theorem 1.2. *There is a procedure that divides* 5 *muffins evenly among* 3 *students such that the smallest piece is* $\frac{5}{12}$. *There is no procedure that yields a larger smallest piece.*

Proof. **Part One: There is a procedure with smallest piece $\frac{5}{12}$.** The following procedure divides and distributes 5 muffins to 3 students such that (1) everyone gets $\frac{5}{3}$ and (2) the smallest piece is $\frac{5}{12}$:

(1) Divide 4 muffins $\{\frac{5}{12}, \frac{7}{12}\}$. (This means that 4 muffins are cut into two pieces, one of size $\frac{7}{12}$, and one of size $\frac{5}{12}$. We will formalize this notation later.)
(2) Divide 1 muffin $\{\frac{6}{12}, \frac{6}{12}\}$.
(3) Give 2 students $\{\frac{6}{12}, \frac{7}{12}, \frac{7}{12}\}$. (This means that 2 students get a piece of size $\frac{6}{12}$ and two pieces of size $\frac{7}{12}$. We will formalize this notation later.)
(4) Give 1 student $\{\frac{5}{12}, \frac{5}{12}, \frac{5}{12}, \frac{5}{12}\}$.

Part Two: Every procedure has a piece $\leq \frac{5}{12}$. Assume that there is a procedure for dividing up 5 muffins and distributing the pieces to 3 students such that every student gets $\frac{5}{3}$ muffins. We show that some piece is $\leq \frac{5}{12}$.

We first modify the procedure. If some muffin is uncut then we modify the procedure to cut that muffin $\{\frac{1}{2}, \frac{1}{2}\}$ and give both halves to the intended recipient of the original uncut muffin. We leave it to the reader to show that the original procedure and our modification of it have the same sized smallest piece.

We can now assume that every muffin is cut into at least 2 pieces.

Case 1: Some muffin is cut into ≥ 3 pieces. Then some piece is $\leq \frac{1}{3}$. Note that $\frac{1}{3} < \frac{5}{12}$. Hence, we have a piece, that is, $< \frac{5}{12}$, so this case is done. Combined with our modification of the procedure we can now assume every muffin is cut into exactly two pieces.

Case 2: Some student gets ≤ 2 pieces. The two pieces add up to $\frac{5}{3}$, hence some piece is $\geq \frac{5}{3} \times \frac{1}{2} = \frac{5}{6}$. Oh. How can a *big piece* help us? Recall that each muffin is cut into two pieces. Look at the piece of size $\geq \frac{5}{6}$. Look at the muffin it came from. The *other piece of that muffin* is $\leq 1 - \frac{5}{6} = \frac{1}{6}$. Note that $\frac{1}{6} < \frac{5}{12}$. Hence, we have a piece, that is, $< \frac{5}{12}$, so this case is done.

Case 3: All 5 muffins are cut into 2 pieces. Hence there are 10 pieces distributed to 3 students. If every student got ≤ 3 pieces, that's only 9 pieces, so that can't happen. Hence some student gets ≥ 4 pieces. So some student has $\frac{5}{3}$ distributed among 4 pieces. One of those pieces is $\leq \frac{5}{3} \times \frac{1}{4} = \frac{5}{12}$. □

Notation 1.3. Note the box at the end of the proof. It is our end-of-proof sign. We use it throughout the book.

1.2. Some Conventions for the Rest of the Book

Before proceeding, we need some conventions that we use throughout the book.

(1) Rather than say *one of the students* we often use Alice or Bob. This makes the proofs and procedures less wordy and more personal.

(2) We say a muffin is cut into *pieces*, and a student gets *shares*; however, pieces and shares are the same thing.

(3) We identify a piece of a muffin with the size of that piece. For example, we use the following phrases:

- *Divide a muffin* $\{\frac{1}{4}, \frac{1}{4}, \frac{1}{2}\}$.
- *Give a student* $\{\frac{5}{12}, \frac{5}{12}, \frac{1}{2}, \frac{7}{12}\}$.
- *We assume all of the pieces are in the closed interval* $[\frac{5}{12}, \frac{7}{12}]$.
- *If there is a piece x then it came from some muffin. The rest of that muffin is* $1 - x$.

1.3. Three Muffins, Five Students

We just solved the *Five Muffins, Three Students* problem. Now let's solve the *Three Muffins, Five Students* problem.

You have 3 muffins and 5 students. You want to divide the muffins evenly so that each student gets $\frac{3}{5}$ muffins. The following picture shows one way to do this:

- RED gets $\frac{3}{5}$ of the first muffin.
- BLUE gets $\frac{3}{5}$ of the second muffin.
- GREEN gets $\frac{3}{5}$ of the third muffin.
- YELLOW gets $\frac{2}{5}$ of the first muffin and $\frac{1}{5}$ of the third muffin.
- BLACK gets $\frac{2}{5}$ of the second muffin and $\frac{1}{5}$ of the third muffin.

For both YELLOW and BLACK, one of the pieces they receive is $\frac{1}{5}$. They are outraged! They think $\frac{1}{5}$ is small since they have big hands.

Exercise 1.4. Is there a way to divide 3 muffins into pieces, and give the pieces to 5 students, such that (1) every student gets $\frac{3}{5}$, and (2) every piece is *bigger* than $\frac{1}{5}$?

Try to solve it yourself before looking at the solution.

Solution to Exercise 1.4

The following picture shows how to divide 3 muffins for 5 students such that (1) everyone gets $\frac{3}{5}$ and (2) the smallest piece is $\frac{1}{4}$:

- RED gets $\frac{6}{20}$ of the first and second muffin.
- BLUE gets $\frac{7}{20}$ of the first muffin and $\frac{5}{20}$ of the third muffin.
- GREEN gets $\frac{7}{20}$ of the first muffin and $\frac{5}{20}$ of the third muffin.
- YELLOW gets $\frac{7}{20}$ of the second muffin and $\frac{5}{20}$ of the third muffin.
- BLACK gets $\frac{7}{20}$ of the second muffin and $\frac{5}{20}$ of the third muffin.

End of Solution to Exercise 1.4

Is there a procedure where the smallest piece is $> \frac{1}{4}$? What do you think is true? What do you hope is true? We show that there is no procedure where the smallest piece is $> \frac{1}{4}$. We also include a formal description of the procedure with smallest piece $\frac{1}{4}$.

Theorem 1.5. *There is a procedure that divides 3 muffins evenly among 5 students such that the smallest piece is $\frac{1}{4}$. There is no procedure that yields a larger smallest piece.*

Proof. **Part One:** There is a procedure with smallest piece $\frac{1}{4}$.

The following procedure divides and distributes 3 muffins to 5 students such that (1) everyone gets $\frac{3}{5}$ and (2) the smallest piece is $\frac{1}{4}$:

(1) Divide 2 muffins $\{\frac{6}{20}, \frac{7}{20}, \frac{7}{20}\}$.
(2) Divide 1 muffin $\{\frac{5}{20}, \frac{5}{20}, \frac{5}{20}, \frac{5}{20}\}$.
(3) Give 4 students $\{\frac{5}{20}, \frac{7}{20}\}$.
(4) Give 1 student $\{\frac{6}{20}, \frac{6}{20}\}$.

Part Two: Every procedure has a piece $\leq \frac{1}{4}$.

Assume that there is a procedure for dividing up 3 muffins and distributing the shares to 5 students. We show that some share is $\leq \frac{1}{4}$.

Case 1: Alice gets ≥ 3 shares. Then some share is $\leq \frac{3}{5} \times \frac{1}{3} = \frac{1}{5} < \frac{1}{4}$.

Case 2: Alice gets 1 share. Then it is of size $\frac{3}{5}$. The muffin that piece came from has $\frac{2}{5}$ left which may or may not itself be cut. There are two cases:

Case 2a: The $\frac{2}{5}$ left is cut into ≥ 2 pieces. Then there is a piece $\leq \frac{2}{5} \times \frac{1}{2} = \frac{1}{5} < \frac{1}{4}$.

Case 2b: The $\frac{2}{5}$ is left intact. Bob gets it. Bob must also get other pieces (perhaps just 1) that add up to $\frac{1}{5}$. Hence some piece is $\leq \frac{1}{5} < \frac{1}{4}$.

Case 3: All 5 students get 2 shares. Hence there are 10 shares. Some muffin is cut into ≥ 4 pieces (if all muffins are cut into ≤ 3 pieces then there are $\leq 9 < 10$ pieces). One of those pieces is $\leq 1 \times \frac{1}{4} = \frac{1}{4}$. □

Exercise 1.6 (Due to Yunseo Choi and Kevin Cong). Finish up the following alternative proof that the *Three-Muffins–Five-student problem* requires some piece to be $\leq \frac{1}{4}$: Assume you have some procedure. If some muffin is cut into ≥ 4 pieces then clearly some piece is $\leq \frac{1}{4}$. Hence you can assume that every muffin is cut into ≤ 3 pieces. Therefore there are ≤ 9 pieces. Determine the most and least shares that a person can get and that will lead to some piece being $\leq \frac{1}{4}$. You fill in the details.

1.4. Floors and Ceilings and Buddies

We introduce notations and easy observations that would have made the proofs of Theorems 1.2 and 1.5 more compact and, more importantly, will be used throughout this book.

1.4.1. *Floors and Ceilings*

Definition 1.7. Let $x \geq 0$.

(1) *The floor of x*, written $\lfloor x \rfloor$, is x rounded down. For example, $\lfloor 3.9 \rfloor = 3$. Note that $\lfloor 3 \rfloor = 3$.
(2) *The ceiling of x*, written $\lceil x \rceil$, is x rounded up. For example, $\lceil 3.1 \rceil = 4$. Note that $\lceil 3 \rceil = 3$.

In the proof of Theorem 1.2 we noted that if 10 shares are given to 3 students then someone gets ≥ 4 shares. This is a case of

the Generalized Pigeon Hole Principle.

We state this principle, in terms of muffins and students, and leave the proof to the reader.

Notation 1.8. $\mathbb{N}$ is the set $\{0, 1, 2, \ldots\}$. See Appendix A for this and other examples of notations for sets of numbers.

Lemma 1.9. *Let $x, s \in \mathbb{N}$. If x shares are given to s students then the following occurs:*

(1) *Some student gets*

$$\geq \left\lceil \frac{x}{s} \right\rceil \text{ shares.}$$

(2) *Some student gets*

$$\leq \left\lfloor \frac{x}{s} \right\rfloor \text{ shares.}$$

Recall Theorem 1.2, Case 2:

> **Case 2:** *All 5 muffins are cut into 2 pieces. Hence there are 10 pieces in total. Alice gets $\geq$ 4 shares (if everyone got $\leq$ 3 shares then there would be $\leq 9 < 10$ pieces). One of those shares is $\leq \frac{5}{3} \times \frac{1}{4} = \frac{5}{12}$.*

We can write this more compactly:

> **Case 2:** *All 5 muffins are cut into 2 pieces. Hence there are 10 shares. Therefore, by Lemma 1.9, Alice gets $\geq \lceil \frac{10}{3} \rceil = 4$ shares. One of those shares is $\leq \frac{5}{3} \times \frac{1}{4} = \frac{5}{12}$.*

In the future we will use Lemma 1.9 without comment.

1.4.2. Intervals

The following is a standard notation and also elaborated on in Appendix A.

Notation 1.10. If $a < b$ are reals then (a, b) is the set of all reals strictly in b

$$(a, b) = \{x : a < x < b\}.$$

Notation 1.11. We will use (a, b) differently. We will often use (a, b) to refer to the set of shares that are in (a, b). So we might say

$$\text{there are 5 shares in } (\tfrac{1}{2}, \tfrac{3}{4}).$$

Keep this in mind while reading the next section on Buddies.

1.4.3. Buddies

Assume that there is a procedure where every muffin is cut into exactly 2 pieces. Let x be one of those pieces. Obviously x came from some muffin. Look at the other piece from that muffin. It is $1 - x$. We call that other piece *the buddy of x*. We formalize this concept.

Definition 1.12. Assume that there is a procedure where every muffin is cut into exactly 2 pieces. Let x be a piece. *The buddy of x*, denoted $B(x)$, is $1 - x$. This definition extends naturally to sets of pieces. Recall from Notation 1.11 that we take (a, b) to mean the set of shares in that interval. Keeping that in mind, the buddy of (a, b) is $(1 - b, 1 - a)$. We denote this $B(a, b)$ rather than the more proper $B((a, b))$. Similarly we use $B[a, b]$. Note that B is a bijection (see Appendix A for the definition of a bijection).

We leave the proof of the following lemma to the reader.

Lemma 1.13. *Assume that there is a procedure where every muffin is cut into two pieces. If (x, y) has c shares in it then $B(x, y) = (1 - y, 1 - x)$ has c shares in it. The same holds for $[x, y]$.*

Recall that in Section 1.1 we had the phrase:

> *Alice cannot get ≤ 2 shares: If she does then there is a share $\geq \tfrac{5}{3} \times \tfrac{1}{2} = \tfrac{5}{6}$. Oh. How can a big share help us? Look at the muffin that big share came from. The other piece of that muffin is $\leq 1 - \tfrac{5}{6} = \tfrac{1}{6} < \tfrac{5}{12}$.*

We can now write this more compactly:

> *Alice cannot get ≤ 2 shares: If she does then there is a share $\geq \tfrac{5}{3} \times \tfrac{1}{2} = \tfrac{5}{6}$. Its buddy is $\leq 1 - \tfrac{5}{6} = \tfrac{1}{6} < \tfrac{5}{12}$.*

1.5. A Meta Problem

The reasoning for the *Five-Muffins, Three-Students* problem, and the *Three-Muffins, Five-Students* problem were strikingly similar. We place the procedures side by side:

5 muffins, 3 students	3 muffins, 5 students
(1) Divide 4 muffins into $\{\frac{5}{12}, \frac{7}{12}\}$.	(1) Divide 2 muffins $\{\frac{6}{20}, \frac{7}{20}, \frac{7}{20}\}$.
(2) Divide 1 muffin into $\{\frac{6}{12}, \frac{6}{12}\}$.	(2) Divide 1 muffin $\{\frac{5}{20}, \frac{5}{20}, \frac{5}{20}, \frac{5}{20}\}$.
(3) Give 2 students $\{\frac{6}{12}, \frac{7}{12}, \frac{7}{12}\}$.	(3) Give 4 students $\{\frac{5}{20}, \frac{7}{20}\}$.
(4) Give 1 student $\{\frac{5}{12}, \frac{5}{12}, \frac{5}{12}, \frac{5}{12}\}$.	(4) Give 1 student $\{\frac{6}{20}, \frac{6}{20}\}$.

Notice the following example of a disturbing duality:

- In the 5-muffins, 3-student procedure, we give 2 students $\{\frac{6}{12}, \frac{7}{12}, \frac{7}{12}\}$.
- In the 3-muffins, 5-students procedure, we divide 2 muffins $\{\frac{6}{20}, \frac{7}{20}, \frac{7}{20}\}$.

Is there a general connection between the x-Muffins, y-Students problem and the y-Muffins, x-Students problem? **Spoiler Alert: Yes.** We will return to this point in Section 2.4.

Chapter 2

One Student! Two Students! Some Basic Theorems!

2.1. Definitions and Notation

We now define the main problem formally.

Definition 2.1. Let $m, s \in \mathbb{N}$. An (m, s)-*procedure* is a procedure to cut m muffins into pieces and then distribute them to the s students so that each student gets $\frac{m}{s}$ muffins. An (m, s)-procedure is *optimal* if it maximizes the size of the smallest piece of any procedure. Let $f(m, s)$ be the size of the smallest piece in an optimal (m, s)-procedure.

Note the following:

- Theorem 1.2 can be restated as $f(5, 3) = \frac{5}{12}$.
- Theorem 1.5 can be restated as $f(3, 5) = \frac{1}{4}$.

It is not obvious that $f(m, s)$ exists. It is not obvious that $f(m, s)$ is always rational. Maybe it involves π (or pie). Is there a program that will, on input $m, s \in \mathbb{N}$, output the rational $f(m, s)$? Note that we could not even ask this question until we knew that $f(m, s)$ exists and is rational. It is not obvious that such a program exists. However, the answer to all three questions is **Yes**. The proofs that $f(m, s)$ exists, is rational, and computable, are in Appendix C.

Note 2.2. Let $m, s \in \mathbb{N}$ and $\alpha \in \mathbb{R}$.

(1) $f(m, s) \geq \alpha$ means that there is an (m, s)-procedure with smallest piece $\geq \alpha$.
(2) $f(m, s) \leq \alpha$ means that every (m, s)-procedure has a piece $\leq \alpha$.

2.2. Easy Facts About f

We will need some facts about f. Here are some exercises to *motivate* them.

Exercise 2.3.

(1) What is $f(21, 7)$?
(2) What is $f(7, 21)$?
(3) Recall from Theorem 1.2 that $f(5, 3) = \frac{5}{12}$. Use both this fact, and the proof of this fact, to determine $f(10, 6)$.

Solution to Exercise 2.3

(1) $f(21, 7) \geq 1$. Procedure: Give all 7 students 3 muffins. No muffin is cut so the smallest piece is of size 1. $f(21, 7) \leq 1$. Actually, for all m, s, $f(m, s) \leq 1$ since there can't be any piece bigger than an entire muffin.
(2) $f(7, 21) = \frac{1}{3}$. Procedure: Cut every muffin $\{\frac{1}{3}, \frac{1}{3}, \frac{1}{3}\}$ and give everyone $\frac{1}{3}$. Clearly can't do any better than $\frac{7}{21} = \frac{1}{3}$.
(3) We look at $f(10, 6)$.

Part One: The Upper Bound
To show $f(10, 6) \leq \frac{5}{12}$, we mimic the proof that $f(5, 3) \leq \frac{5}{12}$.
Assume that there is a procedure for dividing up 10 muffins and distributing the shares to 6 students such that every student gets $\frac{10}{6} = \frac{5}{3}$ muffins. As in the proof of Theorem 1.2 (where we proved $f(5, 3) \leq \frac{5}{12}$) we can assume that no muffin is uncut.

Case 1: Some muffin is split into ≥ 3 pieces. Then some piece is $\leq \frac{1}{3} < \frac{5}{12}$.

Case 2: All muffins are cut into 2 pieces, so there are 20 pieces. Alice gets ≥ 4 shares. By Lemma 1.9 one of those shares is $\leq \frac{2}{3} \times \frac{1}{4} = \frac{5}{12}$.

Part Two: The Lower Bound
We show $f(10, 6) \geq \frac{5}{12}$ by just using $f(5, 3) \geq \frac{5}{12}$. We do not need to use any details of the actual procedure.

Let M_1 be the first 5 muffins and M_2 be the next 5 muffins. Let S_1 be the first 3 students and S_2 be the next 3 students. Use the procedure that showed $f(5, 3) \geq \frac{5}{12}$ on M_1 and S_1. Then use the procedure that showed $f(5, 3) \geq \frac{5}{12}$ on M_2 and S_2. Everyone gets $\frac{5}{3} = \frac{10}{6}$ and the smallest piece is $\frac{5}{12}$.

End of Solution to Exercise 2.3

The solution to Exercise 2.3 inspired the following theorem, whose proof we leave to the reader:

Theorem 2.4. *Let $k, m, s \in \mathbb{N}$.*

(1) *s divides m if and only if $f(m, s) = 1$.*
(2) *m divides s if and only if $f(m, s) = \frac{m}{s}$.*
(3) *For all $k \in \mathbb{N}$, $f(m, s) \leq f(km, ks)$.*
(4) *If s does not divide m then $f(m, s) \leq \frac{1}{2}$.*
(5) *$f(m, s) \geq \frac{1}{s}$.*

Note that Theorem 2.4 *does not* have

$$\text{For all } k \in \mathbb{N}, \ f(km, ks) = f(m, s).$$

However, this is true:

Theorem 2.5. *For all $k, m, s \in \mathbb{N}$, $f(km, ks) = f(m, s)$.*

This was proven by Richard Chatwin (2019). The proof is beyond the scope of this book.

Theorem 2.6. *Let $m, s \in \mathbb{N}$.*

(1) *Assume s does not divide m. If there is an (m, s)-procedure with smallest piece α then there is an (m, s)-procedure with smallest piece α where every muffin is cut into ≥ 2 pieces.*
(2) *Assume s does not divide m. If there is an (m, s)-procedure with smallest piece $\alpha > \frac{1}{3}$ then there is an (m, s)-procedure with smallest piece α where every muffin is cut into 2 pieces.*

Proof. (1) By Theorem 2.4, since s does not divide m, $f(m,s) < 1$. Clearly $f(m,s) \le \frac{1}{2}$ since if there is a piece of size x there is a piece of size $1-x$. Hence if we have a new protocol with additional pieces of size $\frac{1}{2}$ that will not affect the size of the smallest piece.

If Alice got an uncut muffin then modify the procedure to have that muffin cut $\{\frac{1}{2}, \frac{1}{2}\}$ and give both pieces to Alice. Do this for all students who got an uncut muffin.

(2) By Part 1 there is an (m,s)-procedure where every muffin is cut into ≥ 2 pieces. If any muffin was cut into ≥ 3 pieces then there would be a piece of size $\le \frac{1}{3} < \alpha$, which contradicts the premise. $\square$

2.3. One Student! Two Students!

The following theorem, whose proof we leave to the reader, takes care of the $s=1$ and $s=2$ cases:

Theorem 2.7. *For all $m \ge 1$ the following conditions hold*:

(1) $f(m,1) = 1$.
(2) *If m is even, then* $f(m,2) = 1$.
(3) *If m is odd, then* $f(m,2) = \frac{1}{2}$.

2.4. The Duality Theorem

We will relate $f(a,b)$ to $f(b,a)$. But first an example.

Theorem 2.8. $f(14, 5) \ge \frac{11}{25}$.

Proof. We give a $(14, 5)$-procedure.

(1) Divide 10 muffins $\{\frac{11}{25}, \frac{14}{25}\}$.
(2) Divide 4 muffins $\{\frac{12}{25}, \frac{13}{25}\}$.
(3) Give 2 students $\{\frac{14}{25}, \frac{14}{25}, \frac{14}{25}, \frac{14}{25}\}$.
(4) Give 1 student $\{\frac{11}{25}, \frac{11}{25}, \frac{11}{25}, \frac{11}{25}, \frac{13}{25}, \frac{13}{25}\}$.
(5) Give 2 students $\{\frac{11}{25}, \frac{11}{25}, \frac{11}{25}, \frac{12}{25}, \frac{12}{25}, \frac{13}{25}\}$.

$\square$

In Section 1.5 the (3, 5) procedure looks like the (5, 3) procedure but with the muffins and students swapped. We just showed a procedure for $f(14, 5)$ with smallest piece $\frac{11}{25}$. Can we use this to get a procedure for $f(5, 14)$? We swap muffins and students and see what happens. We take the last instruction and make it into the first instruction.

Give 2 students $\{\frac{11}{25}, \frac{11}{25}, \frac{11}{25}, \frac{12}{25}, \frac{12}{25}, \frac{13}{25}\}$
becomes:

Divide 2 muffins $\{\frac{11}{25}, \frac{11}{25}, \frac{11}{25}, \frac{12}{25}, \frac{12}{25}, \frac{13}{25}\}$.

OH — that won't work since the numbers do not add up to 1. AH — let's scale them so that they do. Since they add up to $\frac{14}{5}$, let's multiply each number by $\frac{5}{14}$ to get

Divide 2 muffins $\{\frac{11}{70}, \frac{11}{70}, \frac{11}{70}, \frac{12}{70}, \frac{12}{70}, \frac{13}{70}\}$.

Similarly, since we are going to replace muffin-cutting with student-giving, and each muffin is worth 1 and each student gets $\frac{5}{14}$, we multiply by $\frac{5}{14}$. Hence we have the following proof that $f(5, 14) \geq \frac{11}{70}$:

(1) Divide 2 muffins $\{\frac{11}{70}, \frac{11}{70}, \frac{11}{70}, \frac{12}{70}, \frac{12}{70}, \frac{13}{70}\}$.
(2) Divide 1 muffin $\{\frac{11}{70}, \frac{11}{70}, \frac{11}{70}, \frac{11}{70}, \frac{13}{70}, \frac{13}{70}\}$.
(3) Divide 2 muffins $\{\frac{14}{70}, \frac{14}{70}, \frac{14}{70}, \frac{14}{70}, \frac{14}{70}\}$.
(4) Give 10 students $\{\frac{11}{70}, \frac{14}{70}\}$.
(5) Give 4 students $\{\frac{12}{70}, \frac{13}{70}\}$.

We now present the Duality Theorem which was discovered independently by Erich Friedman and ourselves. We will rarely refer to it again explicitly; however, because of it, we only need to consider the case of $m \geq s$.

Theorem 2.9. *Let $m, s \in \mathbb{N}$. Then $f(s, m) = \frac{s}{m} f(m, s)$.*

Proof. Assume we have the optimal (m, s)-procedure. Note that the smallest piece will be of size $f(m, s)$. We show how to transform it into an (s, m)-procedure with smallest piece $\frac{s}{m} f(m, s)$.

- Replace every instruction of the form

$$\text{Give } t \text{ students } \{a_1, \ldots, a_L\}$$

with

$$\text{Divide } t \text{ muffins } \{\tfrac{a_1 s}{m}, \ldots, \tfrac{a_L s}{m}\}.$$

- Replace every instruction of the form

$$\text{Divide } t \text{ muffins } \{b_1, \ldots, b_L\}$$

with

$$\text{Give } t \text{ students } \{\tfrac{b_1 s}{m}, \ldots, \tfrac{b_L s}{m}\}.$$

We leave it to the reader to prove that applying these rules yields an (s, m)-procedure with smallest piece $\tfrac{s}{m} f(m, s)$.

We have shown (with your help) that $f(s, m) \geq \tfrac{s}{m} f(m, s)$. By a change of variables we get: $f(m, s) \geq \tfrac{m}{s} f(s, m)$.

Hence

$$f(s, m) \geq \frac{s}{m} f(m, s) \geq \frac{s}{m} \times \frac{m}{s} f(s, m) = f(s, m).$$

Therefore $f(s, m) = \tfrac{s}{m} f(m, s)$. □

Chapter 3

Our Plan

We would like an algorithm that will, given m and s, output:

(1) $f(m, s)$.
(2) An (m, s)-procedure with smallest piece $f(m, s)$.
(3) A proof that there is no better procedure.

We do not achieve this; however, we do develop many interesting techniques and determine many $f(m, s)$ (more on that later). We illustrate the structure of most of the chapters by presenting the structure of Chapter 6.

(1) Our techniques so far do not suffice to determine $f(11, 5)$.
(2) We present a new technique to determine $f(11, 5)$.
(3) The new technique determines other unknown values of $f(m, s)$.
(4) We measure our progress.

We measure our progress as follows: Let A be the set of all (m, s) with $3 \leq s \leq 100$, $s < m \leq 110$, m, s relatively prime. (This choice of A pushes our computing resources to the limit.) A has 3520 pairs. Our first general technique for upper bounds will be the Floor–Ceiling method. We later assert that the Floor–Ceiling method gives the correct value of $f(m, s)$ for 2301 of these pairs. This was found by a computer program. More generally, for each method, we will assert for how many pairs in A it was able to determine the correct value of $f(m, s)$ that had not been known by prior methods. This is our measure of progress.

Chapters 4, 6, 8, and 9 present a sequence of methods to obtain upper bounds on $f(m, s)$: FC (Floor–Ceiling), Half, INT (Interval), and MID (Midpoint). These four techniques build on each other.

Chapter 5 presents an algorithm FINDPROC (for Find Procedure) that will, given m, s, α, try to find an (m, s)-procedure with smallest piece $\geq \alpha$, hence *verifying* that $f(m, s) \geq \alpha$. These algorithms suggest a technique to find $f(m, s)$: First find the best upper bound α using FC, Half, INT, and MID, and then run FINDPROC on (m, s, α) to (hopefully) verify that α is a lower bound. If so then we know that $f(m, s) = \alpha$. This determines $f(m, s)$ for around 88% of the ordered pairs (m, s) in A.

So what cases are not covered? Many (though not all) of the cases not covered are when $\lceil \frac{2m}{s} \rceil = 3$. Chapters 10 and 11 present the EBM (easy buddy–match) and HBM (hard buddy–match) methods. They only work when $\lceil \frac{2m}{s} \rceil = 3$. Using FC, Half, INT, MID, EBM, HBM to get an upper bound, and then FINDPROC to verify that it's a lower bound, we determine $f(m, s)$ for around 93% of the ordered pairs in A.

Chapter 12 presents the Gap method, which uses both techniques from the MID method and the buddy–match methods. The Gap method solves *almost* all the remaining cases in A. The Train method, which takes care of the remaining cases in A, is mentioned; however, it is not in the book. It is on the MUFFINS website. In Chapter 12, we present an algorithm that uses all of the methods above and, for *all* of the ordered pairs in A, determines $f(m, s)$. So... are we done? Will our methods solve *every* muffin problem? No. There are cases, not in A, where our methods do not suffice.

Chapter 13 presents the wonderful method of Scott Huddleston and Richard Chatwin which solves every muffin problem quickly. We do not present the proof of correctness.

Chapter 4

Three Students! Four Students! The Floor–Ceiling Theorem!

4.1. Goals for This Chapter

We determine, for all m, $f(m, 3)$ and $f(m, 4)$. On the way, we will obtain a general theorem, *The Floor–Ceiling Theorem*, which is used throughout the book.

4.2. Three Students, Four–Five–Six–Seven Muffins

Exercise 4.1. For $4 \leq m \leq 7$ find $f(m, 3)$.

Solution to Exercise 4.1
In all of the solutions below except $f(6, 3)$ we have that s does not divide m and that we are trying to prove an upper bound $\geq \frac{1}{3}$. Hence, by Theorem 2.6.2, we can assume that any procedure, except the one for $(6, 3)$, cuts every muffin into two pieces.

(1) $f(4, 3) = \frac{1}{3}$: $f(4, 3) \geq \frac{1}{3}$ by Theorem 2.4. Assume, by way of contradiction, that we have a $(4, 3)$-procedure with smallest piece $> \frac{1}{3}$. By Theorem 2.6, we can assume every muffin is cut into two pieces, so there are 8 pieces.

- If Alice gets ≥ 4 shares, then there is a share $\leq \frac{4}{3} \times \frac{1}{4} = \frac{1}{3}$.
- If Alice gets ≤ 2 shares, then there is a share $\geq \frac{4}{3} \times \frac{1}{2} = \frac{2}{3}$ whose buddy is $\leq 1 - \frac{2}{3} = \frac{1}{3}$.
- If everyone has 3 shares, then there are 9 shares. This contradicts that there are 8 shares.

(2) $f(5, 3) = \frac{5}{12}$. This is Theorem 1.2. Recall that there is a student Alice who gets ≥ 4 shares, so some piece is $\leq \frac{5}{3} \times \frac{1}{4} = \frac{5}{12}$.

(3) $f(6, 3) = 1$ is easy.

(4) $f(7, 3) = \frac{5}{12}$.

Here is a procedure for $f(7, 3) \geq \frac{5}{12}$.

(a) Divide 4 muffins $\{\frac{5}{12}, \frac{7}{12}\}$.
(b) Divide 3 muffins $\{\frac{6}{12}, \frac{6}{12}\}$.
(c) Give 2 students $\{\frac{5}{12}, \frac{5}{12}, \frac{6}{12}, \frac{6}{12}, \frac{6}{12}\}$.
(d) Give 1 student $\{\frac{7}{12}, \frac{7}{12}, \frac{7}{12}, \frac{7}{12}\}$.

We show that $f(7, 3) \leq \frac{5}{12}$.

Assume, by way of contradiction, that there is a $(7, 3)$-procedure with smallest piece $> \frac{5}{12}$. By Theorem 2.6, we can assume that every muffin is cut into 2 pieces. Hence there are 14 pieces, so Bob gets $\leq \lfloor \frac{14}{3} \rfloor = 4$ shares. Some share is $\geq \frac{7}{3} \times \frac{1}{4} = \frac{7}{12}$. Its buddy is $\leq 1 - \frac{7}{12} = \frac{5}{12}$.

4.3. The Floor–Ceiling Theorem

The proof that $f(5, 3) \leq \frac{5}{12}$ involved looking at Alice who had many shares. The proof that $f(7, 3) \leq \frac{5}{12}$ involved looking at Bob who had few shares. In the Floor–Ceiling Theorem, we use both a student who has many shares and a student who has few shares.

Theorem 4.2 (The Floor–Ceiling Theorem). *Let $m, s \in \mathbb{N}$, s does not divide m, and $s \leq 2m$. Then*

$$f(m, s) \leq \max\left\{\frac{1}{3}, \min\left\{\frac{m}{s} \times \frac{1}{\lceil \frac{2m}{s} \rceil}, 1 - \frac{m}{s} \times \frac{1}{\lfloor \frac{2m}{s} \rfloor}\right\}\right\}.$$

(The condition $s \leq 2m$ prevents the denominator $\lfloor \frac{2m}{s} \rfloor$ from ever being 0.)

Proof. Assume that there is an (m, s)-procedure. Since s does not divide m, by Theorem 2.6 all muffins are cut into ≥ 2 pieces. So there are $2m$ pieces.

Case 1: If some muffin is cut into ≥ 3 pieces, then there is a piece $\leq \frac{1}{3}$.

Case 2: All muffins are cut into 2 pieces. The following conditions occur:

- Alice gets $\geq \lceil \frac{2m}{s} \rceil$ shares. Hence some share is $\leq \frac{m}{s} \times \frac{1}{\lceil \frac{2m}{s} \rceil}$.
- Bob gets $\leq \lfloor \frac{2m}{s} \rfloor$ shares. Hence some share is $\geq \frac{m}{s} \times \frac{1}{\lfloor \frac{2m}{s} \rfloor}$. Its buddy is $\leq 1 - \frac{m}{s} \times \frac{1}{\lfloor \frac{2m}{s} \rfloor}$.

The result follows. $\square$

We define the function FC (Floor–Ceiling) to encompass both Theorem 4.2 and the trivial case where s divides m. This will make later exposition smoother without special cases.

Notation 4.3. Assume $s \leq m$.

(1) If s divides m, then $FC(m, s) = 1$.
(2) If s does not divide m, then

$$FC(m, s) = \max\left\{\frac{1}{3}, \min\left\{\frac{m}{s} \times \frac{1}{\lceil \frac{2m}{s} \rceil}, 1 - \frac{m}{s} \times \frac{1}{\lfloor \frac{2m}{s} \rfloor}\right\}\right\}.$$

By the Floor–Ceiling Theorem, $f(m, s) \leq FC(m, s)$. How good a bound is this? Informally, $f(m, s) = FC(m, s)$ most of the time (when $m \geq s$). Formally, we have the following theorem.

Theorem 4.4. *If $s \geq 3$, $m \geq s$, and $m \geq \frac{s^3 + 2s^2 + s}{2}$, then $f(m, s) = FC(m, s)$.*

The proof of this theorem is beyond the scope of this book; however, there is a proof on the MUFFIN website.

The FC theorem cannot give an upper bound, that is, $< \frac{1}{3}$. Are there any (m, s) with $m \geq s$ such that $f(m, s) < \frac{1}{3}$? No.

Theorem 4.5. *For all $m \geq s$, $f(m, s) \geq \frac{1}{3}$.*

There is one proof on the MUFFIN website and a different proof due to Richard Chatwin (2019) in his paper.

Theorems 4.4 and 4.5 indicate that $f(m, s) = \mathrm{FC}(m, s)$ happens quite often. Is it the case that, for all $m \geq s$, $f(m, s) = \mathrm{FC}(m, s)$? No. If the answer had been Yes, then this book would have been shorter. Many chapters of the book begin with an (m, s) where $f(m, s) < \mathrm{FC}(m, s)$ and devise a new technique for upper bounds.

Definition 4.6. Fix s. If $f(m, s) < \mathrm{FC}(m, s)$, then we call *m an exception*. The value of s is always understood. For example, if we had the sentence *11 is an exception*, then a prior sentence will have made it clear that we are considering (say) $s = 5$.

On the MUFFIN website there is an extensive analysis of the exceptions. We summarize the results of it in Section 4.9.

4.4. Notation for a Student Having Many Shares

We will need the following notation for the next exercise and throughout this book.

Notation 4.7. If we want to give Alice L_1 shares of size a_1 and L_2 shares of size a_2, and L_3 shares of size a_3 we denote this by

$$\text{Give Alice } \{L_1 : a_1 \ || \ L_2 : a_2 \ || \ L_3 : a_3\}.$$

4.5. Three Students

Note the following (if you do not know mod notation, then see Appendix A).

- If $m \equiv 0 \pmod{3}$, then there exists k such that $m = 3k$.
- If $m \equiv 1 \pmod{3}$, then there exists k such that $m = 3k + 1$.
- If $m \equiv 2 \pmod{3}$, then there exists k such that $m = 3k + 2$.

Exercise 4.8.

(1) Compute $\mathrm{FC}(m, 3)$ for $m = 4, 7, 10, 13, 16, \ldots$ and try to spot a pattern so that you get a formula for $\mathrm{FC}(m, 3)$ when $m \equiv 1 \pmod{3}$. Prove that, for all $m \equiv 1 \pmod{3}$ with $m \geq 4$, $f(m, 3) = \mathrm{FC}(m, 3)$. This

will entail finding procedures. (*Hint*: Get procedures for (4, 3), (7, 3), (10, 3), and try to spot a pattern.)
(2) Compute $FC(m, 3)$ for $m = 5, 8, 11, 14, 17, \ldots$ and try to spot a pattern so that you get a formula for $FC(m, 3)$ when $m \equiv 2 \pmod{3}$. Prove that, for all $m \equiv 2 \pmod{3}$ with $m \geq 5$, $f(m, 3) = FC(m, 3)$. This will entail finding procedures. (*Hint*: Get procedures for (5, 3), (8, 3), (11, 3), and try to spot a pattern.)
(3) State and prove a theorem that gives, for every m, $f(m, 3)$.

Solution to Exercise 4.8
We just do part (3). We give the theorem and the proof together.

Theorem 4.9.
Case 0: $m \equiv 0 \pmod{3}$. If $k \geq 1$, then clearly $f(3k, 3) = 1$.

For the rest of the cases the upper bound is obtained by the Floor–Ceiling Theorem. Hence we just give the procedure.

Case 1: $m \equiv 1 \pmod{3}$. If $k \geq 1$, $f(3k + 1, 3) = \frac{3k-1}{6k}$:

(1) Divide $2k$ muffins $\left\{\frac{3k-1}{6k}, \frac{3k+1}{6k}\right\}$.
(2) Divide $(k + 1)$ muffins $\left\{\frac{1}{2}, \frac{1}{2}\right\}$.
(3) Give 1 student $\left\{2k : \frac{3k+1}{6k}\right\}$.
(4) Give 2 students $\left\{k : \frac{3k-1}{6k} \parallel k + 1 : \frac{1}{2}\right\}$.

Case 2: $m \equiv 2 \pmod{3}$. If $k \geq 1$, $f(3k + 2, 3) = \frac{3k+2}{6k+6}$:

(1) Divide $2k + 2$ muffins $\left\{\frac{3k+2}{6k+6}, \frac{3k+4}{6k+6}\right\}$.
(2) Divide k muffins $\left\{\frac{1}{2}, \frac{1}{2}\right\}$.
(3) Give 1 student $\left\{2k + 2 : \frac{3k+2}{6k+6}\right\}$.
(4) Give 2 students $\left\{k + 1 : \frac{3k+4}{6k+6} \parallel k : \frac{1}{2}\right\}$.

4.6. Four Students

Note the following:

- If $m \equiv 0 \pmod{4}$, then there exists k such that $m = 4k$.
- If $m \equiv 1 \pmod{4}$, then there exists k such that $m = 4k + 1$.

- If $m \equiv 2 \pmod{4}$, then there exists k such that $m = 4k + 2$.
- If $m \equiv 3 \pmod{4}$, then there exists k such that $m = 4k + 3$.

Exercise 4.10. State and prove a theorem that gives, for every m, $f(m, 4)$. (*Hint*: There are four cases: $m \equiv 0, 1, 2, 3 \pmod{4}$.)

Solution to Exercise 4.10
We present the theorem and proof together.

Theorem 4.11. *Cases 1 and 3 both use the upper bound from the Floor–Ceiling Theorem. Hence, in those cases, we only present the lower bound (the procedure).*

Case 0: $m \equiv 0 \pmod{4}$. *If $k \geq 1$, then clearly $f(4k, 4) = 1$.*

Case 1: $m \equiv 1 \pmod{4}$. *If $k \geq 1$,*

(1) *Divide $4k$ muffins $\{\frac{4k-1}{8k}, \frac{4k+1}{8k}\}$.*
(2) *Divide 1 muffin $\{\frac{1}{2}, \frac{1}{2}\}$.*
(3) *Give 2 students $\{2k : \frac{4k-1}{8k} \parallel \frac{1}{2}\}$.*
(4) *Give 2 students $\{2k : \frac{4k+1}{8k}\}$.*

Case 2: $m \equiv 2 \pmod{4}$. *If $k \geq 1$, then clearly $f(4k + 2, 4) = \frac{1}{2}$.*

Case 3: $m \equiv 3 \pmod{4}$. *If $k \geq 1$, $f(4k + 3, 4) = \frac{4k+1}{8k+4}$:*

(1) *Divide $4k + 2$ muffins $\{\frac{4k+1}{8k+4}, \frac{4k+3}{8k+4}\}$.*
(2) *Divide 1 muffin $\{\frac{1}{2}, \frac{1}{2}\}$.*
(3) *Give 2 students $\{2k + 1 : \frac{4k+3}{8k+4}\}$.*
(4) *Give 2 students $\{2k + 1 : \frac{4k+1}{8k+4} \parallel \frac{1}{2}\}$.*

4.7. The Floor–Ceiling Theorem When $m < s$

The Floor–Ceiling Theorem has a condition $s \leq 2m$. We only use it when $s \leq m$ since we only every look at the case $s \leq m$. There is a version of the Floor–Ceiling Theorem that works when $m \geq 2s$. Had this book been written to only deal with the $m \leq s$ case, we would have used this version.

We present it but leave the proof to the reader. *Hint*: Look at the proof of Theorem 1.5.

Theorem 4.12. *If $m, s \in \mathbb{N}$, m does not divide s, and $m \leq 2s$, then*

$$f(m, s) \leq \max\left\{\frac{m}{s} \times \frac{1}{3}, 1 - \frac{m}{s}, \min\left\{\frac{1}{\lceil\frac{2s}{m}\rceil}, \frac{m}{s} - \frac{1}{\lfloor\frac{2s}{m}\rfloor}\right\}\right\}.$$

4.8. Progress Report

We will provide statistics on how well the Floor–Ceiling method works later, in Section 5.9.

4.9. Exceptions

On the MUFFIN website, we have extensive data about exceptions. We use the values of $f(m, s)$ where $3 \leq s \leq 100$ and $s < m \leq 110$ (the same values specified in Chapter 3). We are confident that if we used more data we would get similar results since the patterns revealed themselves early on and persisted.

What Is the Maximum Exception? By Theorem 4.4, we know that, for all s, there are no exceptions $m \geq \frac{s^3 + 2s^2 + s}{2}$. Can this bound be lowered? The empirical evidence indicates the following conditions:

(1) If $s \equiv 1 \pmod{2}$, then the maximum exception is

$$\leq 0.63s^2 - 8.31.$$

(2) If $s \equiv 0 \pmod{4}$, then the maximum exception is

$$\leq 0.16s^2 - 0.03s - 3.16.$$

(3) If $s \equiv 2 \pmod{4}$, then the maximum exception is

$$\leq 0.31s^2 + 0.11s - 5.33.$$

Conjecture 4.13. *There is a constant c such that, for all s, for all $m \geq cs^2$, $f(m, s) = \text{FC}(m, s)$.*

How Many Exceptions Are There?

We have no theorem on the number of exceptions; however, we do not think there are many of them. The empirical evidence indicates the following conditions:

(1) If $s \equiv 1 \pmod 2$ and $s \geq 9$, then the number of exceptions will be
$$\leq 2s - 17.08.$$

(2) If $s \equiv 0 \pmod 4$ and $s \geq 16$, then the number of exceptions will be
$$\leq 1.21s - 16.10.$$

(3) If $s \equiv 2 \pmod 4$ and $s \geq 14$, then the number of exceptions will be
$$\leq 1.75s - 18.75.$$

Conjecture 4.14. *There is a constant c such that, for all s, the number of exceptions is $\leq cs$.*

Where Are the Exceptions?

We have no theorems about where the exceptions are. Looking at limited data Alan Frank noticed that *most* of the exceptions tended to be in two regions: (1) when $m \pmod s$ is *around* $0.25s$, and (2) when $m \pmod s$ is *around* $0.75s$. The empirical evidence indicates he was correct.

Since we do not quantify the notion of *most* or *around* we refrain from making a conjecture.

Chapter 5

Finding Procedures*

5.1. Recap and Goals for This Chapter

If $s \le m$, then $f(m, s) \le \mathrm{FC}(m, s)$. But what about lower bounds? In Chapters 1, 2 and 4, we obtained some lower bounds for $f(m, s)$ by finding (m, s)-procedures. However, the procedures were obtained for particular cases. The methods used there do not extend to other cases.

In this chapter, we do the following procedure:

(1) Find a $(5, 3)$-procedure for $f(5, 3) \ge \frac{5}{12}$ systematically.
(2) Find a $(13, 5)$-procedure for $f(13, 5) \ge \frac{13}{30}$ systematically.
(3) Find a $(17, 15)$-procedure for $f(17, 15) \ge \frac{7}{20}$ systematically. This one has an unusual aspect to it.
(4) Give an algorithm FINDPROC that, given m, s, α, *tries* to find an (m, s)-procedure with smallest piece $\ge \alpha$ to verify $f(m, s) \ge \alpha$.
(5) Clarify what *try* means and what we know about the algorithm's correctness.
(6) Present a program that uses FINDPROC and the Floor–Ceiling Theorem to try to find $f(m, s)$. We then report progress on how many $f(m, s)$'s we can compute from our target set of pairs (see Chapter 3).

See Appendix A for information about sets, multisets, and subsets.

Definition 5.1. Let $A = \{a_1 < a_2 < \cdots < a_L\}$ and let S be a multisubset of A (so an element of A can appear many times in S). The *vector* *representation* of S is a vector $(n_1, n_2, \ldots, n_L)$ where n_i is the number of times that the a_i appears in S.

Notation 5.2. We will often call a multisubset of A just a subset of A.

Definition 5.3. The *sum of a multiset* simply means the sum of the elements in the multiset.

Example 5.4. Let $A = \{19, 20, 21, 22, 23, 24\}$

(1) The multiset

$$\{19, 19, 20\}$$

is a subset of A. We represent it by the vector

$$(2, 1, 0, 0, 0, 0).$$

It sums to $19 + 19 + 20 = 58$.

(2) The multiset

$$\{20, 20, 20, 20, 20, 20, 20, 20, 20, 20, 20\}$$

is a subset of A. We represent it by the vector

$$(0, 11, 0, 0, 0, 0).$$

It sums to $20 \times 11 = 220$.

(3) The multiset

$$\{20, 21, 22, 23, 24\}$$

is a subset of A. We represent it by the vector

$$(0, 1, 1, 1, 1, 1).$$

It sums to

$$20 + 21 + 22 + 23 + 24 = 110.$$

Finding Procedures 29

5.2. Example of Systematic Lower Bounds

The reader is urged to look up systems of linear equations. The keyword *matrix* will help.

In Theorem 1.2 we showed $f(5, 3) \geq \frac{5}{12}$ by using intuition and guesswork to find a $(5, 3)$-procedure. Similar for $f(13, 5) \geq \frac{13}{30}$. In this section, we derive the procedures for both systematically. We also consider $f(17, 15) \geq \frac{7}{20}$ which has an unusual feature.

We will need the following definition.

Definition 5.5. If $S(x_1, \ldots, x_n)$ is a system of linear equations, then *an $\mathbb{N}$-solution to S* is a vector $(a_1, \ldots, a_n) \in \mathbb{N}^n$ such that $S(a_1, \ldots, a_n)$ is satisfied.

5.2.1. $f(5, 3) \geq \frac{5}{12}$ *Systematically*

(1) Suppose we know that the only piece sizes we need for this procedure are

$$A = \left\{ \frac{5}{12}, \frac{6}{12}, \frac{7}{12} \right\}.$$

Each muffin will be split into two piece sizes from A. We can represent this with a multiset of piece sizes. For example, if a muffin is split into two halves, then we may represent it with the multiset $\{\frac{6}{12}, \frac{6}{12}\}$ or with the vector representation $(0, 2, 0)$.

We leave it to the reader to show that every student gets 3 or 4 shares.

In order to avoid cluttering the page with denominators, we will multiply all piece sizes by 12. So instead, let

$$B = \{5, 6, 7\}.$$

If we want to know all ways to cut a muffin into 2 pieces rather than say

We need all 2-subsets of A that sum to 1

we say

We need all 2-subsets of B that sum to $1 \times 12 = 12$

If we want to know all ways to give Alice her share, rather than say

> We need all 3-subsets and 4-subsets of A that sum to $\frac{5}{3}$

we say

> We need all 3-subsets and 4-subsets of B that sum to $\frac{5}{3} \times 12 = 20$

(2) Find all vectors that correspond to how a muffin can be cut. We need all subsets of B that sum to 12 (since $1 = \frac{12}{12}$). While this is easy to guess we do it systematically. Let x_5 be the number of 5's, x_6 be the number of 6's and x_7 be the number of 7's. Then we need all $\mathbb{N}$-solutions to the system of equations

$$5x_5 + 6x_6 + 7x_7 = 12,$$
$$x_5 + x_6 + x_7 = 2.$$

By multiplying the second equation by 5 and subtracting, we get

$$x_6 + 2x_7 = 2$$

We use this to get that the only $\mathbb{N}$-solutions are $(x_6, x_7) = (2, 0)$ and $(x_6, x_7) = (0, 1)$ to obtain that the only $\mathbb{N}$-solutions of the original set of equations is

- $(x_5, x_6, x_7) = (0, 2, 0)$;
- $(x_5, x_6, x_7) = (1, 0, 1)$.

These translate to the following ways to cut a muffin:

- $\{6, 6\}$ which is $(0, 2, 0)$. Let m_1 be the number of muffins cut this way.
- $\{5, 7\}$ which is $(1, 0, 1)$. Let m_2 be the number of muffins cut this way.

(3) Find all vectors that correspond to a student getting 3 shares (we later do 4 shares).

Let x_i be the number of shares of size i. The ways a student can get 3 shares can be found by solving this system of equations:

$$5x_5 + 6x_6 + 7x_7 = 20$$

$$x_5 + x_6 + x_7 = 3$$

The ways a student can get 4 shares can be found by solving a modified system of equations where 3 is replaced by 4.

We leave it to the reader to solve both sets of equations and find the following are the only ways a student can get shares:

- $\{6, 7, 7\}$ which is $(0, 1, 2)$. Let s_1 be the number of students who get these shares.
- $\{5, 5, 5, 5\}$ which is $(4, 0, 0)$. Let s_2 be the number of students who get these shares.

(4) Set up equations to find m_1, m_2, s_1, s_2.

The number of each piece size that the muffins give is equal to the number of each share size that the students receive. Therefore, we get the equation:

$$m_1(0, 2, 0) + m_2(1, 0, 1) = s_1(0, 1, 2) + s_2(4, 0, 0)$$

This equation implies:

$$m_2 = 4s_2,$$

$$2m_1 = s_1,$$

$$m_2 = 2s_1.$$

Since there are 5 muffins and 3 students:

$$m_1 + m_2 = 5,$$

$$s_1 + s_2 = 3.$$

(5) The 5 equations have one N-solution: $m_1 = 1, m_2 = 4, s_1 = 2, s_2 = 1$.

(6) Take the N-solutions and make a procedure out of it.

(1) ($m_1 = 1$) Divide 1 muffin $\{\frac{6}{12}, \frac{6}{12}\}$.
(2) ($m_2 = 4$) Divide 4 muffins $\{\frac{5}{12}, \frac{7}{12}\}$.
(3) ($s_1 = 2$) Give 2 students $\{\frac{6}{12}, \frac{7}{12}, \frac{7}{12}\}$.
(4) ($s_1 = 1$) Give 1 student $\{\frac{5}{12}, \frac{5}{12}, \frac{5}{12}, \frac{5}{12}\}$.

5.2.2. $f(13, 5) \geq \frac{13}{30}$ **Systematically**

In Section 1.1, we took you through our general approach in the case of $f(5, 3) \geq \frac{5}{12}$. In this section, *you* will prove that $f(13, 5) \geq \frac{13}{30}$.

Exercise 5.6. In this exercise you will show $f(13, 5) \geq \frac{13}{30}$. Assume that there is a $(13, 5)$-procedure. Assume that:

- All of the pieces are in

$$A = \left\{\frac{13}{30}, \frac{14}{30}, \frac{15}{30}, \frac{16}{30}, \frac{17}{30}\right\}.$$

- By Theorem 2.6, all of the muffins are cut into two pieces.

(1) Find all subsets of piece sizes that correspond to how a muffin can be cut. Represent them as vectors.
(2) Find all subsets of piece sizes that correspond to what a student can get. Represent them as vectors.
(3) Set up equations like the ones in Section 5.2.1 which will equate the muffin-view with the student-view.
(4) Solve those equations (use a software package).
(5) Give a procedure for $f(13, 5) \geq \frac{13}{30}$ using the last part.

Solution to Exercise 5.6
We clear fractions. The pieces are in

$$B = \{13, 14, 15, 16, 17\}.$$

A muffin is of size $1 \times 30 = 30$, and each students gets $\frac{13}{5} \times 30 = 78$.

Finding Procedures

(1) Find all vectors that correspond to how a muffin can be cut. We need all 2-subsets of B that sum to 30. Hence we need all $\mathbb{N}$-solutions to:

$$13x_{13} + 14x_{14} + 15x_{15} + 16x_{16} + 17x_{17} = 30,$$

$$x_{13} + x_{14} + x_{15} + x_{16} + x_{17} = 2.$$

Here they are:

- $\{13, 17\}$, which is $(1, 0, 0, 0, 1)$. Let m_1 be the number of muffins cut this way.
- $\{14, 16\}$, which is $(0, 1, 0, 1, 0)$. Let m_2 be the number of muffins cut this way.
- $\{15, 15\}$, which is $(0, 0, 2, 0, 0)$. Let m_3 be the number of muffins cut this way.

(2) Find all vectors that correspond to what a student can get. We need all subsets of B that sum to 78.

We leave it to the reader to show that Alice can't get ≥ 6 pieces or ≤ 4 pieces. Hence we need 5-subsets and 6-subsets of B that sum to 78. Hence we need to all $\mathbb{N}$-solutions to

$$13x_{13} + 14x_{14} + 15x_{15} + 16x_{16} + 17x_{17} = 78,$$

$$x_{13} + x_{14} + x_{15} + x_{16} + x_{17} = 5$$

and to

$$13x_{13} + 14x_{14} + 15x_{15} + 16x_{16} + 17x_{17} = 78,$$

$$x_{13} + x_{14} + x_{15} + x_{16} + x_{17} = 6.$$

Here they are:

- $\{13, 13, 13, 13, 13, 13\}$ which is $(6, 0, 0, 0, 0)$. Let s_1 be the number of students who get these shares.
- $\{15, 15, 16, 16, 16\}$ which is $(0, 0, 2, 3, 0)$. Let s_2 be the number of students who get these shares.
- $\{15, 15, 15, 16, 17\}$ which is $(0, 0, 3, 1, 1)$. Let s_3 be the number of students who get these shares.

- $\{14, 16, 16, 16, 16\}$ which is $(0, 1, 0, 4, 0)$. Let s_4 be the number of students who get these shares.
- $\{14, 15, 16, 16, 17\}$ which is $(0, 1, 1, 2, 1)$. Let s_5 be the number of students who get these shares.
- $\{14, 15, 15, 17, 17\}$ which is $(0, 1, 2, 0, 2)$. Let s_6 be the number of students who get these shares.
- $\{14, 14, 16, 17, 17\}$ which is $(0, 2, 0, 1, 2)$. Let s_7 be the number of students who get these shares.
- $\{13, 16, 16, 16, 17\}$ which is $(1, 0, 0, 3, 1)$. Let s_8 be the number of students who get these shares.
- $\{13, 15, 16, 17, 17\}$ which is $(1, 0, 1, 1, 2)$. Let s_9 be the number of students who get these shares.
- $\{13, 14, 17, 17, 17\}$ which is $(1, 1, 0, 0, 3)$. Let s_{10} be the number of students who get these shares.

(4) Equate the muffin pieces with the student shares:

$$m_1(1, 0, 0, 0, 1) + m_2(0, 1, 0, 1, 0) + m_3(0, 0, 2, 0, 0)$$
$$= s_1(6, 0, 0, 0, 0) + s_2(0, 0, 2, 3, 0) + s_3(0, 0, 3, 1, 1) + s_4(0, 1, 0, 4, 0)$$
$$+ s_5(0, 1, 1, 2, 1) + s_6(0, 1, 2, 0, 2) + s_7(0, 2, 0, 1, 2)$$
$$+ s_8(1, 0, 0, 3, 1) + s_9(1, 0, 1, 1, 2) + s_{10}(1, 1, 0, 0, 3).$$

This equation implies:

$$m_1 = 6s_1 + s_8 + s_9 + s_{10},$$
$$m_2 = s_4 + s_5 + s_6 + 2s_7 + s_{10},$$
$$2m_3 = 2s_2 + 3s_3 + s_5 + 2s_6 + s_9,$$
$$m_2 = 3s_2 + s_3 + 4s_4 + 2s_5 + s_7 + 3s_8 + s_9,$$
$$m_1 = s_3 + s_5 + 2s_6 + 2s_7 + s_8 + 2s_9 + 3s_{10}.$$

Since there are 13 muffins and 5 students, we have

$$m_1 + m_2 + m_3 = 13,$$
$$s_1 + s_2 + s_3 + s_4 + s_5 + s_6 + s_7 + s_8 + s_9 + s_{10} = 5.$$

(5) There are 13 ℕ-solutions to the 7 equations. We present 5 of them. We convert one of them to a procedure. The reader is invited to find the other ℕ-solutions and to convert any or all of them to procedures.

Solution One:

- $m_1 = 6, m_2 = 7, m_3 = 0$.
- $s_1 = 1, s_4 = 1, s_7 = 3$, all other s_i's are 0.

This ℕ-solution yields the following procedure:

(1) ($m_1 = 6$) Divide 6 muffins $\left\{\frac{13}{30}, \frac{17}{30}\right\}$.
(2) ($m_2 = 7$) Divide 7 muffins $\left\{\frac{14}{30}, \frac{16}{30}\right\}$.
(3) ($s_1 = 1$) Give 1 student $\left\{\frac{13}{30}, \frac{13}{30}, \frac{13}{30}, \frac{13}{30}, \frac{13}{30}, \frac{13}{30}\right\}$.
(4) ($s_4 = 1$) Give 1 student $\left\{\frac{14}{30}, \frac{16}{30}, \frac{16}{30}, \frac{16}{30}, \frac{16}{30}\right\}$.
(5) ($s_7 = 3$) Give 3 students $\left\{\frac{14}{30}, \frac{14}{30}, \frac{16}{30}, \frac{17}{30}, \frac{17}{30}\right\}$.

Solution Two:

- $m_1 = 6, m_2 = 6, m_3 = 1$.
- $s_1 = 1, s_2 = 1, s_7 = 3$, all other s_i's are 0.

Solution Three:

- $m_1 = 6, m_2 = 5, m_3 = 2$.
- $s_1 = 1, s_2 = 1, s_6 = 1, s_7 = 2$, all other s_i's are 0.

Solution Four:

- $m_1 = 6, m_2 = 4, m_3 = 3$.
- $s_1 = 1, s_2 = 1, s_6 = 2, s_7 = 1$, all other s_i's are 0.

Solution Five:

- $m_1 = 6, m_2 = 3, m_3 = 4$.
- $s_1 = 1, s_2 = 1, s_6 = 3$, all other s_i's are 0.

5.2.3. $f(17, 15) \geq \frac{7}{20}$ Systematically

We want to show that $f(17, 15) \geq \frac{7}{20}$. This is unusual in that the denominator of the answers is not a multiple of s. These cases are rare but they do happen. Should we assume the pieces all have denominator 15? 20? Neither. We assume the pieces all have denominator 60, the least common multiple of 15 and 20. Hence all pieces are in

$$A = \left\{\frac{21}{60}, \frac{22}{60}, \ldots, \frac{39}{60}\right\}.$$

Note that there are 19 elements in A.

Exercise 5.7.

(1) Find all subsets of A that correspond to how a muffin can be cut. These are 2-subsets that sum to 60.
(2) Find all subsets of A that correspond to what a student can get. These are 2-subsets or 3-subsets that sum to 68.
(3) Set up equations like the ones in Section 5.2.1 that equate the muffin-view and the student-view.
(4) Solve those equations (use a software package).
(5) Give a procedure for $f(17, 15) \geq \frac{21}{60}$ using the last part.

5.3. The Algorithm FINDPROC

We describe the algorithm FINDPROC that has the following input–output behavior:

(1) Input: (m, s, α) where α is rational.
(2) Output: Either a procedure that shows $f(m, s) \geq \alpha$ or **DK**. If there is an $\mathbb{N}$-solution then output $f(m, s) \geq \alpha$ and go to the next step. (**DK** stands for Don't Know since we don't know if $f(m, s) \geq \alpha$ (though we have never seen a case where $f(m, s) = \alpha$ and the program outputs **DK**).

Here is a description of the algorithm.

(1) Input(m, s, α). Let $\alpha = \frac{c}{d}$ in lowest terms.
(2) If $\alpha = \frac{1}{3}$ then use Theorem 4.5. Henceforth we assume $\alpha > \frac{1}{3}$.
(3) Let b be the least common multiple of s and d. (Usually $b = d$.)

(4) Let $a = \frac{bc}{d}$. Note that $a \in \mathbb{N}$ since b is a multiple of d. Note that $\alpha = \frac{a}{b}$ and $1 - \alpha = \frac{b-a}{b}$. We assume the pieces are all of sizes:

$$\left\{\frac{a}{b}, \frac{a+1}{b}, \ldots, \frac{b-a}{b}\right\}.$$

(See Conjecture 5.9 for why we can assume it.) We clear fractions and think of the piece sizes as

$$B = \{a, a+1, \ldots, b-a\}.$$

With this in mind (1) every muffin is of size $1 \times b = b$, and (2) every student gets $\frac{m}{s} \times b = \frac{bm}{s}$ (since s divides b, $\frac{bm}{s} \in \mathbb{N}$).

(5) Find all 2-subsets of B that sum to b. Represent them by vectors. Let M be the set of all these vectors. The vectors in M represent all ways to divide a muffin. (We can find these vectors either using equations, as described in Section 5.4, or using "recursion" as described in Section 5.6. You will see why the word is in quotes later.)

(6) Let $V = \left\lceil \frac{2m}{s} \right\rceil$ (in Section 6.9 we discuss this choice). Find all V-subsets and $(V-1)$-subsets of B that sum to $\frac{bm}{s}$. Represent them by vectors. Let S be the set of all these vectors. The vectors in S represent all ways to give shares to students. (We can find these vectors either using equations, as described in Section 5.4, or again using "recursion".)

(7) Find all $\mathbb{N}$-solutions of the following system of linear equations: The variables are

$$\{m_{\vec{v}} : \vec{v} \in M\} \cup \{s_{\vec{u}} : \vec{u} \in S\}.$$

We equate the muffin-viewpoint and the student-viewpoint:

$$\sum_{\vec{v} \in M} m_{\vec{v}} \vec{v} = \sum_{\vec{u} \in S} s_{\vec{u}} \vec{u}.$$

(The equation above yields $|B| = b - 2a + 1$ linear equations.) Since there are m muffins and s students, we have

$$\sum_{\vec{v} \in M} m_{\vec{v}} = m,$$

$$\sum_{\vec{u} \in S} s_{\vec{u}} = s.$$

(8) If there is an N-solution then output $f(m, s) \geq \alpha$ and go to the next step. If not, then output **DK**.
(9) Use the N-solution found to generate a procedure showing $f(m, s) \geq \alpha$. Each $m_{\vec{v}}$ tells the number of muffins which are split into pieces according to the vector $\vec{v}$. Likewise, each $s_{\vec{u}}$ tells the number of students who receive shares according to their vector $\vec{u}$.

5.4. How to Find the Vectors: Equations Approach

There are two ways to find the vectors. We describe one in detail here that we have seen in our examples. We also describe some shortcuts.

Here is the problem we often face: we are given a set

$$B = \{a, a+1, \ldots, b-a\}$$

along with c, $V-1$, V, T_1, T_2 and need to find

- All 2-subsets of B that sum to T_1.
- All $(V-1)$-subsets of B that sum to T_2.
- All V-subsets of B that sum to T_2.

More generally we need all W-subsets of B that sum to T. For $a \leq i \leq b-a$, let x_i be the number of times we use i. Then we need to find all N-solutions to the system

$$ax_a + \cdots + (b-a)x_{b-a} = T,$$

$$x_a + \cdots + x_{b-a} = W.$$

Every N-solution to this system corresponds to a W-subset of B that sums to T. All such W-subsets of B can be obtained this way.

In future chapters we will find that (1) the $(V-1)$-subsets only use the larger elements in B, and (2) the V-subsets only use the smaller elements in B. We may also find other numbers that are forbidden. When coding this up one can use this to get equations with less variables.

5.5. Recursion, Dynamic Programming, and "Recursion"

This section is a brief aside about three techniques one could use for a recurrence. We discuss them since the last technique, "Recursion", will be used in the next section. We give an example that is not connected to muffins.

Consider the recurrence:

$a_0 = 1$;
$a_1 = 5$;
$a_2 = 10$;
$(\forall n \geq 3)[a_n = a_{\lfloor \sqrt{n} \rfloor} + a_{\lfloor 2 \log_2(n) \rfloor - 1}]$.

I want a program that will, on input n, return a_n.

The following program does the problem recursively.

Program A

(1) Input(n).
(2) If $n = 0$ return 1. If $n = 1$ return 5.
(3) Otherwise return $A(\lfloor \sqrt{n} \rfloor) + A(\lfloor 2\log_2(n) \rfloor - 1)$.

The program is elegant but it computes more than it needs to. Consider what happens when $n = 1000$.

- The program calls $A(31)$ and $A(17)$.
- The call of $A(31)$ calls $A(5)$ and $A(7)$.
- The call of $A(17)$ calls $A(4)$ and $A(7)$.

We stop here and make our point. Note that $A(7)$ was called twice. More generally, there may be a lot of unneeded recomputation.

Hence we look at another approach, called *Dynamic Programming*. We do the recursion bottom up rather than top down.

Program A

(1) Input(n). We have an array A.
(2) $A[0] \leftarrow 1$. If $n = 0$ return $A[0]$.
(3) $A[1] \leftarrow 5$. If $n = 1$ return $A[1]$.
(4) $A[2] \leftarrow 10$. If $n = 2$ return $A[2]$.

(5) For $i = 3$ to n $A[i] \leftarrow A\left[\lfloor\sqrt{i}\rfloor\right] + A\left[\lfloor 2\log_2(i) - 1\rfloor\right]$.
(6) Return $A[n]$.

The key to this program is that when you are computing (say) $A[18]$, you have all the prior $A[i]$'s that you need. Also note that in the end you have $A[0], \ldots, A[n]$, more than you want. Hence we did some unneeded computation.

Is there a way to do this problem with less unneeded computation? Yes. We combine the two techniques. We keep an array of already-computed values. The array is global. It exists and is updated through all calls to the function. Whenever we call the function we first check if the answer is already in the array. If so we return it.

Initialize Array

(1) $A[0] \leftarrow 1$.
(2) $A[1] \leftarrow 5$.
(3) $A[2] \leftarrow 10$.

Program A

(1) Input(n). If $n = 0$ or $n = 1$ or $n - 2$ then return $A[n]$.
(2) Check if $A[n]$ is defined. If so then return it.
(3) Return $A\left(\lfloor\sqrt{n}\rfloor\right) + A\left(\lfloor 2\log_2(n) - 1\rfloor\right)$ and put it in the array. (Recall that these calls to A first checked to see if the value was already known.)

We call this last approach "Recursion". In the literature it is called *memoization*. This name is so unintuitive that I refuse to use it. Hence I use "Recursion".

We will use "Recursion" to find the vectors.

5.6. How to Find the Vectors: "Recursion" Approach

We solve a more general problem than the one we need.

Definition 5.8. Let $B = \{x_1, \ldots, x_L\}$, $T \in \mathbb{Z}$, and $k \in \mathbb{N}$. We assume all $x_i \geq 1$. Let $F(i, T, k)$ be the set of all k-subsets of $\{x_1, \ldots, x_i\}$ that sum to T. Note that $F(0, T, k)$ is looking at subsets of $\emptyset$ (the empty set) that sum to T. (The case of $T \leq -1$ is an edge case that we rarely need.)

How will $F(i, T, k)$ help us? Henceforth let

$$B = \{a, a+1, \ldots, b-a\}.$$

For the algorithm FINDPROC we need to compute the following:

- The 2-subsets of $\{a, \ldots, b-1\}$ that sum to b. This is $F(b-2a+1, b, 2)$.
- The V-subsets and $(V-1)$-subsets of $\{a, \ldots, b-a\}$ that sum to $\frac{bm}{s}$. This is $F(b-2a+1, \frac{bm}{s}, V-1) \cup F(b-2a+1, \frac{bm}{s}, V)$.

We will end up computing more values of $F(i, T, k)$ then we need; however, this turns out to be the best way to do it.

Below you will see that sometimes we have $F(i, T, k) = \emptyset$ and sometimes $F(i, T, k) = \{\emptyset\}$. This is not a typo. We describe the difference:

- $F(i, T, k) = \emptyset$ means that there are no subsets of

$$\{a, \ldots, a+i-1\}$$

of size k that sum to T.
- $F(i, T, k) = \{\emptyset\}$ means that the empty set is a subset of

$$\{a, \ldots, a+i-1\}$$

of size k that sums to T. (In this case $T=0$ and $k=0$.)

(1) For all i, with $i \geq 1$, $F(i, 0, 0) = \{\emptyset\}$. The sum of the elements of $\emptyset$ is 0.
(2) For all i, for all $k \geq 1$, $T(i, 0, k) = \emptyset$. There are no k-subsets that sum to 0 since $k \geq 1$.
(3) For all T, k with $T \geq 1$, $F(0, T, k) = \emptyset$. There are no k-subsets of $\emptyset$ that sum to T.
(4) For all T, i with $T \geq 1$, $F(i, T, 0) = \emptyset$. There are no 0-subsets that sum to T.
(5) For all i, T, k with $T \leq -1$, $F(i, T, k) = \emptyset$. There are no i-subsets of $\{x_1, \ldots, x_i\}$ that sum to a negative number. We will need this in a recursion.

(6) For all k, $F(1, kx_1, k) = \{\{x_1, \ldots, x_1\}\}$ (there are k x_1's). We seek all k-subset of $\{x_1\}$ that sums to kx_1. There only one is $\{x_1, \ldots, x_1\}$ (k x_1's).
(7) For all k, for all $T \neq kx_1$, $F(1, T, k) = \emptyset$. We seek a k-subset of $\{x_1\}$ that sums to T. There is only one k-subset of $\{x_1\}$ which is $\{x_1, \ldots, x_1\}$ (k times). This subset does not sum to T. Hence there is no such k-subset.
(8) Let $1 \leq j \leq i \leq L$ and $F(i, x_j, 1) = \{\{x_j\}\}$. We seek a 1-subset of $\{x_1, \ldots, x_i\}$ that sums to x_j. The only such set is $\{x_j\}$.
(9) Let $1 \leq i \leq L$, $T \notin \{x_1, \ldots, x_i\}$ and let $F(i, T, 1) = \emptyset$. We seek a 1-subset of $\{x_1, \ldots, x_i\}$ that sums to T. Since $T \notin \{x_1, \ldots, x_i\}$, there is no such set.

The first case not covered is $F(2, T, 2)$. This will reduce to one of the cases above.

Let $Y \in F(2, T, 2)$. So Y is a 2-subset of $\{x_1, x_2\}$ that sums to T. There are cases depending on if x_2 is in Y or not.

- $x_2 \notin Y$. Then Y is a 2-subset of $\{x_1\}$ that sums to T. It means $Y \in F(1, T, 2)$. Great!
- $x_2 \in Y$. What is $Y - \{x_2\}$? It is a 1-subset of $\{x_1, x_2\}$ that sums to $T - x_2$. It means $Y - \{x_2\} \in F(2, T - x_2, 1)$. Hence Y is of the form $\{x_2\} \cup X$ where $X \in F(1, T - x_2, 1)$. Note that if $x_2 > T$, then $T - x_2 < 0$ and this is taken care of above.

More succinctly:

$$F(2, T, 2) = F(1, T, 2) \cup \{\{x_2\} \cup X \mid X \in F(1, T - x_2, 1)\}.$$

We now generalize. We want to find all $Y \in F(i, T, k)$. There are two cases depending on x_i in Y.

- $x_i \notin Y$. Then Y is a k-subset of $\{x_1, \ldots, x_{i-1}\}$ that sums to T. This means $Y \in F(i - 1, T, k)$. Great!
- $x_i \in Y$. What is $Y - \{x_i\}$? It is a $(k - 1)$-subset of $\{x_1, \ldots, x_k\}$ that sums to $T - x_i$. This means $Y - \{x_i\} \in F(i, T - x_2, k - 1)$. Hence Y is of the form $\{x_i\} \cup X$ where $X \in F(i, T - x_i, k - 1)$. Note that if $x_i > T$, then $T - x_i < 0$ and this is taken care of in the list of easy cases.

More succinctly:

$$F(i, T, k) = F(i - 1, T, k) \cup \{\{x_i\} \cup X \mid X \in F(i, T - x_i, k - 1)\}$$

We give the algorithm for F. Note that while trying to compute $F(i, T, k)$, it will compute other $F(i', T', k')$ along the way.

Let $B = \{x_1, \ldots, x_i\}$ be given. We do not regard it as part of the input.

We will keep array $A(i, T, k)$ in storage of already-known values of $F(i, T, k)$. The idea is that if we are trying to determine $F(i, T, k)$ and need to know some $F(i', T', k')$ with either $i' < i$, $T' < T$ or $k' < k$ then we will either already have that value in storage or we will try to compute it. Hence, the first two things we do are (1) see if i, T, k are an easy case, and (2) check if $F(i, T, k)$ has already been computed and stored.

$F(i, T, k)$

(1) If (i, T, k) is one of the easy cases discussed above, then output the answer noted above.
(2) Check if $A(i, T, k)$ has an answer. If so, then output it.
(3) (If you got to this step then (i, T, k) is not an easy case and is not already known.) Compute $F(i - 1, T, k)$. (This is a recursive call to F. Note that if the answer has already been computed and is in the array A then this step will be fast.)
(4) Compute $F(i, T - x_i, k - 1)$. (This is a recursive call to F. Note that if the answer has already been computed and is in the array A then this step will be fast.)
(5) Output (and put in $A(i, T, k)$)

$$F(i - 1, T, k) \cup \{\{x_i\} \cup X \mid X \in F(i, T - x_i, k - 1)\}.$$

5.7. Comments On Speed

In our experience the equations approach is faster when m is small (say less than 50) and the "recursion" method is faster when m is large. However, both approaches tend to be slow when $m, s \geq 100$. FINDPROC is the main bottleneck in our programs. Either algorithm is exponential in m, s.

5.8. There Is No Try, Only Do

If the algorithm FINDPROC(m, s, α) outputs a procedure, then great! We know $f(m, s) \geq \alpha$ and have a proof of that.

If the algorithm outputs *We could not prove that* $f(m, s) \geq \alpha$ then we *do not* know that there is no (m, s)-procedure with smallest piece α. All we know is that there is no such procedure with denominator

$$b = \text{lcm}\{\text{denominator of } \alpha, s\}.$$

We have never seen a case where FINDPROC($m, s, \frac{c}{d}$) returned **DK** and yet there was an (m, s)-procedure with smallest piece $\geq \frac{c}{d}$. We conjecture that we never will see such a case.

Conjecture 5.9. *If* $f(m, s) = \frac{c}{d}$ *and b is the least common multiple of s and d, then there is a procedure showing* $f(m, s) \geq \frac{c}{d}$ *where all of the pieces are of the form $\frac{i}{b}$ where $i \in \mathbb{N}$.*

5.9. Program and Progress

Using FC and FINDPROC, we have the following attempt at an algorithm to find $f(m, s)$:

(1) Input(m, s).
(2) $\alpha = $ FC(m, s).
(3) Run FINDPROC(m, s, α).
(4) If it outputs a procedure P then output α. Otherwise output **DK** (for Don't Know).

There are 3520 pairs (m, s) we are considering (see Chapter 3).

- For 2301 of them $f(m, s) = $ FC(m, s). This is $\sim 65.37\%$.
- For 1219 of them the functions FC and FINDPROC did not suffice to find $f(m, s)$. This is $\sim 34.63\%$.

Chapter 6

The Half Method

6.1. Recap and Goals for This Chapter

From Chapter 4 we know that for $1 \le s \le m$,

$$f(m, s) \le \text{FC}(m, s).$$

Is it the case that, for all $m \ge s$, $f(m, s) = \text{FC}(m, s)$? No. The counterexample with the smallest s is $f(11, 5)$:

$$f(11, 5) = \frac{13}{30} < \frac{11}{25} = \text{FC}(11, 5).$$

The proof of the upper bound uses a new technique, which we call *The Half method*. We develop an algorithm Half (m, s) which, given m, s outputs α such that $f(m, s) \le \alpha$.

Definition 6.1. Let $m, s, V \in \mathbb{N}$. Assume that there is an (m, s)-procedure.

(1) A V-*student* is a student who gets V shares.
(2) A share that goes to a V-student is a V-*share*.

6.1.1. Five Students, Six, Seven, ..., Thirteen Muffins

Exercise 6.2. For $6 \le m \le 13$, find $f(m, 5)$. *Hint:* First use the Floor–Ceiling Theorem to get an upper bound and then try to show that it is a lower bound. *Warning:* This approach will fail when $m = 11$.

Solution to Exercise 6.2
- $f(6,5) = \frac{2}{5}$:

$f(6,5) \le \frac{2}{5}$ by the Floor–Ceiling Theorem. We leave it to the reader to come up with a procedure.

Hint: Every muffin is divided $\{\frac{2}{5}, \frac{3}{5}\}$.

- $f(7,5) = \frac{1}{3}$:

$f(7,5) \le \frac{1}{3}$ by the Floor–Ceiling Theorem. We leave it to the reader to come up with a procedure.

Hint: Assume that everyone gets either $\{\frac{5}{15}, \frac{8}{15}, \frac{8}{15}\}$ or $\{\frac{7}{15}, \frac{7}{15}, \frac{7}{15}\}$.

- $f(8,5) = \frac{2}{5}$.

$f(8,5) \le \frac{2}{5}$ by the Floor–Ceiling Theorem. We leave it to the reader to come up with a procedure.

Hint: Every muffin is divided $\{\frac{2}{5}, \frac{3}{5}\}$.

- $f(9,5) = \frac{2}{5}$.

$f(9,5) \le \frac{2}{5}$ by the Floor–Ceiling Theorem. We leave it to the reader to come up with a procedure.

Hint: Every muffin is divided $\{\frac{2}{5}, \frac{3}{5}\}$.

- $f(10,5) = 1$. This is easy.

- $f(11,5)$.

By the Floor–Ceiling Theorem $f(11,5) \le \frac{11}{25}$. We will try to prove $f(11,5) \ge \frac{11}{25}$. Let's narrow down what such a procedure will look like. (We could run FINDPROC; however, we prefer to essentially do that by hand and see what happens.)

MUFFINS:

Since $\frac{11}{25} > \frac{1}{3}$, by Theorem 2.6 every muffin is cut into exactly 2 pieces. Hence there are 22 pieces.

STUDENTS:

(1) If Alice has ≥ 6 shares, then one of them is $\le \frac{11}{5} \times \frac{1}{6} = \frac{11}{30} < \frac{11}{25}$.

(2) If Bob has ≤ 3 shares, then one of the shares is $\ge \frac{11}{5} \times \frac{1}{3} = \frac{11}{15}$. Its buddy is $\le 1 - \frac{11}{15} = \frac{4}{15} < \frac{11}{25}$.

The Half Method   47

Hence everyone is either a 4-student or 5-student. Let s_4 (s_5) be the number of 4-students (5-students). The total number of pieces is $4s_4 + 5s_5$. It's also $2 \times 11 = 22$. Hence

$$4s_4 + 5s_5 = 22.$$

The total number of students is 5, hence

$$s_4 + s_5 = 5.$$

These two equations have only one solution: $s_4 = 3$, $s_5 = 2$.

Every student gets $\frac{11}{5} = \frac{11 \times 5}{5 \times 5} = \frac{55}{25}$. We make a leap and assume that every share is in the set

$$A = \left\{\frac{11}{25}, \frac{12}{25}, \frac{13}{25}, \frac{14}{25}\right\}.$$

We list all sets of 4-subsets and 5-subsets of A that sum up to $\frac{55}{25}$. There are only 2:

$$\left\{\frac{13}{25}, \frac{14}{25}, \frac{14}{25}, \frac{14}{25}\right\} \quad \text{and} \quad \left\{\frac{11}{25}, \frac{11}{25}, \frac{11}{25}, \frac{11}{25}, \frac{11}{25}\right\}.$$

Hence the three 3-students get the first set and the two 5-students get the second set. Therefore, the following must be the last two steps of the procedure:

(1) Give 2 students $\{\frac{11}{25}, \frac{11}{25}, \frac{11}{25}, \frac{11}{25}, \frac{11}{25}\}$.
(2) Give 3 students $\{\frac{13}{25}, \frac{14}{25}, \frac{14}{25}, \frac{14}{25}\}$.

Because of the last step there are 12 pieces that are $\geq \frac{13}{25} > \frac{1}{2}$. **This cannot happen.** There were originally 11 muffins, each cut into 2 pieces, so there are at most 11 pieces $> \frac{1}{2}$. What's going on? Is our assumption all of the pieces have denominator 25 wrong? Or is there no (11,5)-procedure with smallest piece $\geq \frac{11}{25}$? In Section 6.2, we will describe a new technique to bound $f(11, 5)$. The bound will be $\frac{13}{30} < \frac{11}{25}$.

- $f(12, 5) = \frac{2}{5}$.

  $f(12, 5) \leq \frac{2}{5}$ by the Floor–Ceiling Theorem. We leave it to the reader to come up with a procedure.

  Hint: Every muffin is divided $\{\frac{2}{5}, \frac{3}{5}\}$.

- $f(13, 5) = \frac{13}{30}$.

By the Floor–Ceiling Theorem $f(13, 5) \le \frac{13}{30}$. We will try to prove $f(13, 5) \ge \frac{13}{30}$. Let's narrow down what such a procedure will look like.

We leave it to the reader to derive that there are four 5-students and one 6-student.

Let's make a leap and assume that every share is in the set

$$A = \left\{\frac{13}{30}, \frac{14}{30}, \frac{15}{30}, \frac{16}{30}, \frac{17}{30}\right\}.$$

There is one 6-student. The only way 6 shares can add up to $\frac{13}{5} = \frac{78}{30}$ is

$$\left\{\frac{13}{30}, \frac{13}{30}, \frac{13}{30}, \frac{13}{30}, \frac{13}{30}, \frac{13}{30}\right\}.$$

Since one student must get this set of shares, there must be 6 muffins cut $\{\frac{13}{30}, \frac{17}{30}\}$. Hence we have the following partial procedure:

(1) Divide 6 muffins $\{\frac{13}{30}, \frac{17}{30}\}$.
(2) Give 1 student $\{\frac{13}{30}, \frac{13}{30}, \frac{13}{30}, \frac{13}{30}, \frac{13}{30}, \frac{13}{30}\}$.

We leave it to the reader to complete the procedure. Start with the fact that there are 6 muffins of size $\frac{17}{30}$ that need to be used.

6.2. A New Technique

By the Floor–Ceiling Theorem we have $f(11, 5) \le \frac{11}{25}$. We proved that there is no (11, 5)-procedure with smallest piece $\ge \frac{11}{25}$ where all of the pieces had denominator 25. We use the ideas in that proof to *derive* a better upper bound.

Theorem 6.3. $f(11, 5) = \frac{13}{30}$.

Proof. We leave the proof that $f(11, 5) \ge \frac{13}{30}$ to the reader. Alternatively, the reader can run FINDPROC(11, 5, $\frac{13}{30}$).

We derive the upper bound during the proof. We will denote it α. Assume, by way of contradiction, that there is an (11, 5)-procedure with every piece $> \alpha$. We assume $\alpha \ge \frac{1}{3}$. By Theorem 2.6, every muffin is cut

The Half Method

into exactly 2 pieces. Hence there are 22 pieces. Note that there can be at most 11 pieces $> \frac{1}{2}$. This will be a key to getting a contradiction.

We follow the lead of the attempt at proving $f(11, 5) \le \frac{11}{25}$ by assuming that everyone is a 4-student or 5-student. But we turn that around: we use that assumption to lower bound α.

Case 1: If Alice gets ≥ 6 shares then some share is

$$\le \frac{11}{5 \times 6} = \frac{11}{30} \le \alpha.$$

We will need $\frac{11}{30} \le \alpha$ to get a contradiction.

Case 2: If Alice gets ≤ 3 shares then some share is

$$\ge \frac{11}{5 \times 3} = \frac{11}{15}.$$

Its buddy is

$$\le 1 - \frac{11}{15} = \frac{4}{15} \le \alpha.$$

We will need $\frac{4}{15} \le \alpha$ to get a contradiction.

Case 3: Everyone is either a 4-student or a 5-student.

Let s_4 (s_5) be the number of 4-students (5-students). Since every muffin is cut into 2 pieces there are $11 \times 2 = 22$ pieces.

Hence,

$$4s_4 + 5s_5 = 22,$$

$$s_4 + s_5 = 5.$$

Hence $s_4 = 3$ and $s_5 = 2$. So there are twelve 4-shares and ten 5-shares. Since there are 11 muffins, each cut in half, there are at most 11 pieces $> \frac{1}{2}$. In particular not all 12 of the 4-shares can be $> \frac{1}{2}$. We will *derive* what α needs to be to ensure that all the 4-shares are $> \frac{1}{2}$. This will be our contradiction.

We want α such that there are no 4-shares $\le \frac{1}{2}$. Assume, by way of contradiction, that there is a 4-share $\le \frac{1}{2}$. The remaining 3 shares add up to $\geq \frac{11}{5} - \frac{1}{2} = \frac{17}{10}$; hence some share is $\geq \frac{17}{10} \times \frac{1}{3} = \frac{17}{30}$. Its buddy is $\leq 1 - \frac{17}{30} = \frac{13}{30}$.

We will need $\frac{13}{30} \leq \alpha$ to get a contradiction.

Putting it all together we need

$$\alpha = \max\left\{\frac{11}{30}, \frac{8}{30}, \frac{13}{30}\right\} = \frac{13}{30}$$

to get a contradiction. We have just proved $f(11, 5) \leq \frac{13}{30}$. □

6.3. The Half Method

In Theorem 6.3 we proved $f(11, 5) \leq \frac{13}{30}$. A brief review:

(1) Since $\frac{13}{30} > \frac{1}{3}$, by Theorem 2.6, every muffin is cut into 2 pieces, so there are $2m$ pieces.
(2) Since each muffin is cut into 2 pieces that are buddies, there are at most 11 pieces that are $> \frac{1}{2}$.
(3) We showed that any procedure with smallest piece $> \frac{13}{30}$ would have at least 12 shares $> \frac{1}{2}$. This gave a contradiction.

We generalize this technique, which we call *The Half method*. It works just as well if we end up with more than m shares $< \frac{1}{2}$.

There are many cases of the Half method. Therefore we give 4 more examples of what can happen when it is applied:

- $f(45, 26) \leq \frac{32}{78}$;
- $f(29, 17) \leq \frac{27}{68}$;
- $f(23, 13) \leq \frac{11}{26}$;
- $f(13, 11) \leq \frac{1}{3}$.

These results may look mysterious; however, after proving them we will speculate on how they could have been derived. This speculation will come to fruition when we describe the algorithm Half.

6.4. $f(45, 26) \le \frac{32}{78}$ by The Half Method

Theorem 6.4. $f(45, 26) = \frac{32}{78}$.

Proof. We leave the proof that $f(45, 26) \ge \frac{32}{78}$ to the reader. Alternatively the reader can run FINDPROC$(45, 26, \frac{32}{78})$.

Assume, by way of contradiction, that there is a $(45, 26)$-procedure with smallest piece $> \frac{32}{78}$. By Theorem 2.6 every muffin is cut into exactly 2 pieces. Hence there are 90 pieces. Note that there can be at most 45 pieces $< \frac{1}{2}$. We show that there is a piece $\le \frac{32}{78}$.

Every student gets $\frac{45}{26} = \frac{45 \times 3}{26 \times 3} = \frac{135}{78}$.

Case 1: Alice gets ≥ 5 shares. Then one of them is $< \frac{135}{78} \times \frac{1}{5} = \frac{27}{78} < \frac{32}{78}$.

Case 2: Bob gets ≤ 2 shares. Then one of the shares is $> \frac{135}{78} \times \frac{1}{2} = \frac{67.5}{78}$. Its buddy is $< 1 - \frac{67.5}{78} = \frac{10.5}{78} < \frac{32}{78}$.

In the subsequent cases we assume the negation of Cases 1 and 2. Hence everyone is either a 3-student or a 4-student. Let s_3 (s_4) be the number of 3-students (4-students). Since there are 90 pieces and 26 students,

$$3s_3 + 4s_4 = 90,$$

$$s_3 + s_4 = 26.$$

Hence $s_3 = 14$ and $s_4 = 12$. So there are fourteen 3-students, twelve 4-students, forty-two 3-shares, and forty-eight 4-shares. Since $48 > 45$, if all of the 4-shares are $< \frac{1}{2}$, that will be a contradiction. Indeed, this will be our contradiction.

We now look at intervals.

Case 3: Alice has a 4-share $\ge \frac{39}{78}$. Alice's other three 4-shares add up to $\le \frac{135}{78} - \frac{39}{78} = \frac{96}{78}$, hence one of them is $\le \frac{96}{78} \times \frac{1}{3} = \frac{32}{78}$.

Case 4: Bob has a 3-share $\le \frac{43}{78}$. Bob's other two 3-shares add up to $\ge \frac{135}{78} - \frac{43}{78} = \frac{92}{78}$, hence one of the shares is $\ge \frac{92}{78} \times \frac{1}{2} = \frac{46}{78}$. Its buddy is $\le 1 - \frac{46}{78} = \frac{32}{78}$.

Case 5: The following picture captures the negation of cases 1, 2, 3 and 4.

$$(\ 48 \ \text{4-shs} \ )[\ 0 \ ](\ 42 \ \text{3-shs} \ )$$
$$\frac{32}{78} \qquad\qquad \frac{39}{78} \ \frac{43}{78} \qquad\qquad \frac{46}{78}$$

The midpoint is $\frac{1}{2} = \frac{39}{78}$. Note that all forty-eight 4-shares are $< \frac{1}{2}$. This is a contradiction. □

We show how one could *derive* the upper bound $f(45, 26) \le \frac{32}{78}$. Let α be the upper bound. We derive conditions on α that will make the proof of $f(45, 26) \le \alpha$ work. We assume $\alpha > \frac{1}{3}$. We guess everyone is either a 3-student or a 4-student. (Later Theorem 6.14 will tell us what to guess.)

In the proof that $f(45, 26) \le \frac{32}{78}$ we deduced that there are forty-two 3-shares and forty-eight 4-shares. This calculation *did not use that the goal was* $\frac{32}{78}$. Hence we can use that reasoning. We have the following picture, though we do not know x or y.

$$(\ 48 \ \text{4-shs} \ )[\ 0 \ ](\ 42 \ \text{3-shs} \ )$$
$$\alpha \qquad\qquad x \quad y \qquad\qquad 1 - \alpha$$

What are x and y?

- x is the least number such that every 4-share is $< x$. Hence $3\alpha + x = \frac{135}{78}$, so $x = \frac{135}{78} - 3\alpha$.
- y is the largest number such that every 3-share is $> y$. Hence $2(1 - \alpha) + y = \frac{135}{78}$, so $y = 2\alpha - \frac{7}{26}$.

Hence we have:

$$(\ 48 \ \text{4-shs} \qquad)[\quad 0 \quad](\quad 42 \ \text{3-shs} \qquad)$$
$$\alpha \qquad\qquad \frac{135}{78} - 3\alpha \quad 2\alpha - \frac{7}{26} \qquad\qquad 1 - \alpha$$

If $x \le \frac{1}{2} \le y$ then there will be 48 > 45 shares to the left of $\frac{1}{2}$ which is a contradiction. We look at setting $x = \frac{1}{2}$ and $y = \frac{1}{2}$.

If $x = \frac{1}{2}$ then

$$\alpha = \frac{\frac{135}{78} - \frac{1}{2}}{3} = \frac{16}{39}.$$

If $y = \frac{1}{2}$ then

$$\alpha = \frac{\frac{1}{2} + \frac{7}{26}}{2} = \frac{5}{13}.$$

You would think we should take the lower value, $\alpha = \frac{5}{13}$. But, alas, if you try to do the proof with this value you get that $y < x$ so the proof would not work. Hence we take $x = \frac{16}{39}$.

6.5. $f(29, 17) \le \frac{27}{68}$ by The Half Method

In the proof of Theorem 6.4, the intervals containing the 3-shares and the intervals containing the 4-shares did not overlap. (This is the most common case for the Half method.) Is there a case where the intervals overlap and the Half method still works? Yes. We present one.

Theorem 6.5. $f(29, 17) = \frac{27}{68}$.

Proof. We leave the proof that $f(29, 27) \ge \frac{27}{68}$ to the reader. Alternatively, the reader can run FINDPROC$(29, 27, \frac{27}{68})$.

Assume, by way of contradiction, that there is a $(29, 17)$-procedure with smallest piece $> \frac{27}{68}$. By Theorem 2.6, every muffin is cut into exactly 2 pieces. Hence there are 58 pieces. Note that there can be at most 29 pieces $> \frac{1}{2}$.

Every student gets $\frac{29}{17} = \frac{29 \times 4}{17 \times 4} = \frac{116}{68}$.

We leave as an exercise to show that (1) if Alice has ≥ 5 shares then she has a share $< \frac{27}{68}$, (2) if Bob has a ≤ 2 shares then one of them has a buddy, that is, $< \frac{27}{68}$, hence (3) everyone is a 3-student or a 4-student, and (4) there are ten 3-students, seven 4-students, thirty 3-shares, and twenty-eight 4-shares. Since $30 > 29$, if all of the 3-shares are $> \frac{1}{2}$, then that will be a contradiction. Indeed, this will be our contradiction.

We now look at intervals.

Case 1: Alice has a 4-share $\geq \frac{35}{68}$. Alice's other three 4-shares sum to $\leq \frac{116}{68} - \frac{35}{68} = \frac{81}{68}$, hence one of them is $\leq \frac{81}{68} \times \frac{1}{3} = \frac{27}{68}$.

Case 2: Bob has a 3-share $\leq \frac{34}{68}$. Bob's other two 3-shares sum to $\geq \frac{116}{68} - \frac{34}{68} = \frac{82}{68}$, hence one of the shares is $\geq \frac{82}{68} \times \frac{1}{2} = \frac{41}{68}$. Its buddy is $\leq 1 - \frac{41}{68} = \frac{27}{68}$.

Case 3: The negation of cases 1 and 2. I know what you are thinking. We'll just draw the picture and have a good sense of what is going on. But the picture is hard to draw. Why? Let's draw the 4-share and 3-share pictures separately.

The 4-shares:

$$(\ 28 \ 4\text{-shs} \ )(\ 0 \ 4\text{-shs} \ )$$
$$\frac{27}{68} \qquad\qquad \frac{35}{68} \qquad\qquad \frac{41}{68}$$

The 3-shares:

$$(\ 0 \ 3\text{-shs} \ )(\ 30 \ 3\text{-shs} \ )$$
$$\frac{27}{68} \qquad\qquad \frac{34}{68} \qquad\qquad \frac{41}{68}$$

They overlap. The interval $(\frac{34}{68}, \frac{35}{68})$ can contain both 3-shares and 4-shares. Can our proof proceed anyway? Yes.

All thirty 3-shares are bigger than $\frac{1}{2}$. This is a contradiction. Hence this case cannot occur. (There may also be some 4-shares in $(\frac{34}{68}, \frac{35}{68})$ but this does not affect the argument.) □

Exercise 6.6. Derive that the upper bound for $f(29, 17)$ using the Half method is $\frac{27}{68}$. (*Hint*: See the paragraphs after the proof of Theorem 6.4.)

6.6. $f(23, 13) \leq \frac{11}{26}$ by The Half Method

In the proof that $f(45, 26) \leq \frac{16}{39}$ we had an interval of 3-shares and an interval of 4-shares. Both were nonempty. There are cases such that when you try to use the Half method, one of the intervals is empty. In our experience, when this happens $f(m, s)$ is smaller than the upper bound given by the Half method (hence other methods are needed to get the correct bound).

Nevertheless, when using the method one must be aware of edge cases. We give an example and discuss how to handle it.

Theorem 6.7. $f(23, 13) \leq \frac{11}{26}$.

Proof. Assume, by way of contradiction, that there is a (23, 13)-procedure with smallest piece $> \frac{11}{26}$. By Theorem 2.6 every muffin is cut into exactly 2 pieces. Hence there are 46 pieces. Note that there can be at most 23 pieces $< \frac{1}{2}$. We show that there is a piece $\leq \frac{11}{26}$.

Every student gets $\frac{23}{13} = \frac{23 \times 2}{13 \times 2} = \frac{46}{26}$.

We leave as an exercise to show that there are six 3-students, seven 4-students, eighteen 3-shares, and twenty-eight 4-shares. Since $28 > 23$, if all of the 4-shares are $< \frac{1}{2}$, that will be a contradiction. Indeed, this will be our contradiction.

We now look at intervals.

4-shares: If Alice has a 4-share $\geq \frac{x}{26}$ (we want to determine x) then her three other 4-shares sum to $\leq \frac{46-x}{26}$, hence one of them is $\frac{(46-x)/3}{26}$. We need to know what values of x will yield a piece, that is, $\leq \frac{11}{26}$. Hence we determine when x satisfies

$$\frac{(46-x)/3}{26} \leq \frac{11}{26}.$$

Any $x \geq 13$ suffice. We take $x = 13$ to minimize the interval of 4-shares. Hence all of Alice's 4-shares are in $(\frac{11}{26}, \frac{13}{26})$.

3-shares: If Bob has 3-shares $\leq \frac{y}{26}$ (we want to determine y) then his two other shares sum to $\geq \frac{46-y}{26}$, hence one of the shares is $\geq \frac{1}{2} \times \frac{46-y}{26}$. Its buddy is $\leq 1 - \frac{1}{2} \times \frac{46-y}{26} = \frac{(y+6)/2}{26}$. We need to know what values of y will yield a piece $\leq \frac{11}{26}$. Hence we determine when y satisfies

$$\frac{(y+6)/2}{26} \leq \frac{11}{26}.$$

Any $y \leq 16$ works. Oh. Since the entire interval is $(\frac{11}{26}, \frac{15}{26})$ and there are no 3-shares $\leq \frac{16}{26}$ there are no 3-shares! We will soon see that this does not affect the Half method; however, it is a sign that the Half method does not yield the best upper bound.

The following picture captures what we know:

$$(\ 28 \ 4\text{-shs} \ )[\ 0 \ ]$$
$$\frac{11}{26} \qquad\qquad \frac{13}{26} \ \frac{15}{26}$$

All twenty-eight 4-shares are smaller than $\frac{1}{2}$. This is a contradiction. Hence this case cannot occur. □

In the proof that $f(23, 13) \le \frac{11}{26}$ the Half method worked since all that was needed is that all the 4-shares are $< \frac{1}{2}$. We did not have to care about the 3-shares.

Another case that could occur (though we've never seen it) is that (say) the 4-shares are the entire interval. This would happen if the right-endpoint of the 4-shares was $\ge \frac{15}{26}$. In this case we would consider the interval of 4-shares to be $(\frac{11}{26}, \frac{15}{26})$.

Exercise 6.8. Prove that $f(23, 13) \le \frac{32}{78}$. (*Hint:* Follow the proof of $f(23, 13) \le \frac{11}{26}$; however, do not do the Half method. Instead use that there are no 3-shares.)

Since $f(23, 13) \le \frac{32}{78}$ we wonder if $f(23, 13) = \frac{32}{78}$. No. In Theorem 9.3 we will show $f(23, 13) \le \frac{53}{130}$ in Chapter 9 using the MID technique. And yes, $f(23, 13) = \frac{53}{130}$.

6.7. $f(13, 11) \le \frac{1}{3}$ by The Half Method

In Theorems 6.4 and 6.7 the right endpoint of the 4-share interval was $\frac{1}{2}$. In Theorem 6.5 the right endpoint of the 3-share interval was $\frac{1}{2}$. In all of these cases it was important where $\frac{1}{2}$ was. Can the Half method work if $\frac{1}{2}$ is not an endpoint? Yes. We give an example.

Theorem 6.9. $f(13, 11) = \frac{1}{3}$.

Proof. We leave the proof that $f(13, 11) \ge \frac{1}{3}$ to the reader. Alternatively the reader can run FINDPROC(11, 5, $\frac{13}{30}$) or use Theorem 4.5.

$f(13, 11) \ge \frac{1}{3}$ by Theorem 4.5. The proof of this theorem is on the MUFFIN website; however, the reader should be able to work out the procedure for $f(13, 11) \ge \frac{1}{3}$ themselves.

The Half Method

Assume, by way of contradiction, that there is a (13, 11)-procedure with smallest piece $> \frac{1}{3}$. By Theorem 2.6 every muffin is cut into exactly 2 pieces. Hence there are 26 pieces. Note that there can be at most 13 pieces $> \frac{1}{2}$. We show that there is a piece $\leq \frac{1}{3}$.

Every student gets $\frac{13}{11} = \frac{13 \times 3}{11 \times 3} = \frac{39}{33}$.

We leave as an exercise to show that there are seven 2-students, four 3-students, fourteen 2-shares, and twelve 3-shares. Since $14 > 13$, if all of the 2-shares are $> \frac{1}{2}$, that will be a contradiction. Indeed, this will be our contradiction.

We leave as an exercise to show that the following picture captures what we know so far:

$$(\text{ 12 3-shs })(\text{ 14 2-shs })$$
$$\frac{11}{33} \qquad\qquad \frac{17}{33} \qquad\qquad \frac{22}{33}$$

All fourteen 2-shares are bigger than $\frac{17}{33} > \frac{16.5}{33} = \frac{1}{2}$. This is a contradiction. Note that the fact that $\frac{1}{2}$ was within the 3-share interval did not matter. □

We show how to derive the upper bound $f(13, 11) \leq \frac{1}{3}$.

Let α be the upper bound. We assume $\alpha \geq \frac{1}{3}$. We guess everyone is either a 2-student or a 3-student. (Later Theorem 6.14 will tell us what to guess.) By the same reasoning used in the proof above there are fourteen 2-shares and twelve 4-shares. So we want α such that all the 2-shares are in $(\frac{1}{2}, 1 - \alpha)$. That will be a contradiction since there are at most 13 shares $> \frac{1}{2}$.

Assume Alice has a 2-share $\leq \frac{1}{2}$. Alice's other share is $\geq \frac{13}{11} - \frac{1}{2} = \frac{15}{22}$. Its buddy is $\leq \frac{7}{22}$. Suppose we take $\alpha = \frac{7}{22}$. OH, that does not work since we need $\alpha \geq \frac{1}{3}$. Or does it work after all?

What we actually showed is that

if $\alpha \geq \frac{7}{22}$ AND every muffin is cut into 2 pieces then there is no procedure with smallest piece $> \alpha$.

This statement is true, but does not give us an upper bound of $\frac{7}{22}$ since a procedure with smallest piece $\frac{7}{22}$ may well have a muffin cut into 3 pieces. But note that (1) $\frac{1}{3} > \frac{7}{22}$ and (2) by Theorem 2.6 if the smallest piece is $> \frac{1}{3}$ then every muffin is cut into two pieces. Hence we have $\alpha = \frac{1}{3}$.

What happened above will be part of our algorithm: If our formula gives a value $< \frac{1}{3}$ then the answer is $\frac{1}{3}$.

6.8. Exercises on Deriving and Verifying

Exercise 6.10.

(1) Prove each of the following using the Half method. *Hint*: You will need a V such that everyone is either a V-student or a $(V - 1)$-student. Try $V = \lceil \frac{2m}{s} \rceil$. *Advice*: There are a lot of problems here. Work on them until they stop being fun.

(a) $f(7, 6) \leq \frac{1}{3}$.
(b) $f(8, 7) \leq \frac{5}{14}$.
(c) $f(19, 7) \leq \frac{25}{56}$.
(d) $f(11, 9) \leq \frac{13}{36}$.
(e) $f(29, 9) \leq \frac{41}{90}$.
(f) $f(38, 9) \leq \frac{59}{126}$.
(g) $f(17, 10) \leq \frac{2}{5}$.
(h) $f(19, 11) \leq \frac{9}{22}$.
(i) $f(41, 11) \leq \frac{61}{132}$.
(j) $f(52, 11) \leq \frac{83}{176}$.
(k) $f(15, 13) \leq \frac{9}{26}$.
(l) $f(22, 13) \leq \frac{21}{52}$.
(m) $f(29, 13) \leq \frac{45}{104}$.
(n) $f(55, 13) \leq \frac{85}{182}$.
(o) $f(19, 16) \leq \frac{1}{3}$.
(p) $f(20, 17) \leq \frac{1}{3}$.
(q) $f(69, 41) \leq \frac{67}{164}$.

(2) For the above exercise pretend the upper bound you can get from the Half method was not given. Derive what it should be. Then see if you are correct.

The Half Method

(3) Write a program that will, given m, s ($m \geq s$ and m, s) determine the smallest α such that $f(m, s) \leq \alpha$ can be proven using the Half method.

Solution to Exercise 6.10
We sketch the solutions to some of the problems by giving the relevant pictures:

- $f(7, 6) \leq \frac{1}{3}$.

$$(\text{ 6 3-shs })(\text{ 8 2-shs })$$
$$\frac{2}{6} \qquad \frac{3}{6} \qquad \frac{4}{6}$$

- $f(8, 7) \leq \frac{5}{14}$.

$$(\text{ 6 3-shs })[\ 0\ ](\text{ 10 2-shs })$$
$$\frac{5}{14} \qquad \frac{6}{14} \quad \frac{7}{14} \qquad \frac{9}{14}$$

- $f(20, 17) \leq \frac{1}{3}$.

$$(\text{ 18 3-shs })(\text{ 22 2-shs })$$
$$\frac{17}{51} \qquad \frac{26}{51} \qquad \frac{34}{51}$$

Note that $\frac{1}{2} = \frac{25.5}{51}$.

- $f(69, 41) \leq \frac{67}{164}$.

$$(\text{ 60 4-shs })[\ 0\ ](\text{ 78 3-shs })$$
$$\frac{67}{164} \qquad \frac{75}{164} \quad \frac{82}{164} \qquad \frac{97}{164}$$

Note 6.11. For every $f(m, s) \leq \alpha$ that you verified in Exercise 6.10 it turns out that $f(m, s) = \alpha$. We leave it to the reader to show this.

6.9. The V-Conjecture

We have the following conjecture.

Conjecture 6.12 (The V-Conjecture). *Let $m \geq s$. Let $V = \lceil \frac{2m}{s} \rceil$. There is an optimal (m, s)-procedure such that everyone is either a V-student or a $(V - 1)$-student.*

Note 6.13. We are not conjecturing that in *every* optimal procedure everyone is either a $(V-1)$-student or a V-student, just that *some* optimal procedure has this property. The stronger statement is false. $f(15, 8) \le \frac{3}{8}$ by the Floor–Ceiling Theorem. Note that $V = \lceil \frac{30}{8} \rceil = 4$. $f(15, 8) \ge \frac{3}{8}$ by the following two procedures, one of which uses 3-student and 4-students but the other one uses 3-students and 5-students. This example is due to Scott Huddleston.

Procedure One

(1) Divide 6 muffins $\{\frac{3}{164}, \frac{5}{164}\}$.
(2) Divide 9 muffins $\{\frac{4}{164}, \frac{4}{164}\}$.
(3) Give 2 students $\{\frac{5}{164}, \frac{5}{164}, \frac{5}{164}\}$.
(4) Give 6 students $\{\frac{3}{164}, \frac{4}{164}, \frac{4}{164}, \frac{4}{164}\}$.

Procedure Two

(1) Divide 15 muffins $\{\frac{3}{164}, \frac{5}{164}\}$.
(2) Give 3 students $\{\frac{5}{164}, \frac{5}{164}, \frac{5}{164}\}$.
(3) Give 5 students $\{\frac{3}{164}, \frac{3}{164}, \frac{3}{164}, \frac{3}{164}, \frac{3}{164}\}$.

Using the V-conjecture we will use the Half method to obtain an α such that $f(m, s) \le \alpha$. We then show $f(m, s) \le \alpha$ without using the V-conjecture.

6.10. Algorithms We Will Need

In the proof of Theorem 6.4 we needed to do the following:

(1) Find the V such that everyone is a $(V-1)$-student or a V-student.
(2) Find s_{V-1}, the number of $(V-1)$-students.
(3) Find s_V, the number of V-students.
(4) Find the interval that contains the $(V-1)$-students.
(5) Find the interval that contains the V-students.

We describe algorithms to find the five items above. We will then use them to describe two programs: (1) VHalf(m, s, α) which will, given

m, s, α, determine if $f(m, s) \leq \alpha$ can be proven with the Half method, and (2) Half(m, s) which will produce the smallest α such that one can show $f(m, s) \leq \alpha$ with the Half method.

6.10.1. An Algorithm that Finds V, s_V, s_{V-1} Given the V-Conjecture

Assume $m > s$, $f(m, s) > \frac{1}{3}$, and s does not divide m. Assume that there is an (m, s)-procedure. By Theorem 2.6, every muffin is cut into exactly 2 pieces. Hence there are $2m$ pieces. We assume that everyone gets either $V - 1$ or V shares, but we do not know what V is. Let s_{V-1} (s_V) be the number of students who get $V - 1$ (V) shares. Hence:

$$(V - 1)s_{V-1} + V s_V = 2m,$$

$$s_{V-1} + s_V = s.$$

The solution is

$$s_{V-1} = Vs - 2m,$$

$$s_V = 2m - s(V - 1).$$

If $V \geq \lceil \frac{2m}{s} \rceil + 1$ then $s_V < 0$. If $V \leq \lceil \frac{2m}{s} \rceil - 1$ then $s_{V-1} < 0$. Hence

- $V = \lceil \frac{2m}{s} \rceil$.
- $s_{V-1} = Vs - 2m$.
- $s_V = 2m - s(V - 1)$.
- The number of $(V - 1)$-shares is $(V - 1)s_{V-1}$.
- The number of V-shares is $V s_V$.

With this in mind we can write an algorithm, named SV, that, given m, s, finds V, s_{V-1}, s_V:

SV(m, s)

(1) Input(m, s).
(2) $V \leftarrow \lceil \frac{2m}{s} \rceil$.
(3) $s_{V-1} \leftarrow Vs - 2m$.
(4) $s_V \leftarrow 2m - s(V - 1)$.

By the above calculations we have the following theorem:

Theorem 6.14. *Assume $m > s$, s does not divide m, and $f(m,s) > \frac{1}{3}$. If the V-conjecture is true, then $V = \lceil \frac{2m}{s} \rceil$.*

6.10.2. An Algorithm that Finds the Intervals

Assume that there is an (m,s)-procedure with smallest piece $> \alpha$. We will assume $\alpha \geq \frac{1}{3}$. From Section 6.10.1, V, s_{V-1}, and s_V can be found. We derive the intervals for the $(V-1)$-shares and the V-shares.

Assume Alice has $V-1$ shares. Let y be the largest number that is smaller than any $(V-1)$-share. Assume, by way of contradiction (actually we will be setting y as large as possible to get a contradiction) that there is a $(V-1)$-share of size y. If y is removed then the $V-2$ shares left add up to $\frac{m}{s} - y$. One of those $V-2$ shares is $\geq \frac{\frac{m}{s} - y}{V-2}$. Its buddy is

$$\leq 1 - \frac{\frac{m}{s} - y}{V - 2}.$$

We get a contradiction if $1 - \frac{\frac{m}{s} - y}{V-2} \leq \alpha$. We get the smallest y by setting:

$$\alpha = 1 - \frac{\frac{m}{s} - y}{V - 2},$$

$$y = \frac{m}{s} - (1 - \alpha)(V - 2).$$

Hence the $(V-1)$-shares are in

$$\left(\frac{m}{s} - (1-\alpha)(V-2),\ 1-\alpha\right).$$

Assume Bob has V shares. Let x be the smallest number that is larger than any V-share. We leave it to the reader to show that

$$x = \frac{m}{s} - \alpha(V - 1).$$

Hence the V-shares are in

$$\left(\alpha,\ \frac{m}{s} - \alpha(V-1)\right).$$

The Half Method

We can now write the algorithm to find these intervals. The algorithm also accounts for strange cases like when x or y is not in the interval.

FINDEND(m, s, α, V)

(1) Input(m, s, α).
(2) $y \leftarrow \frac{m}{s} - (1-\alpha)(V-2)$.
(3) If $y \geq 1 - \alpha$ then $y \leftarrow 1 - \alpha$. If $y \leq \alpha$ then $y \leftarrow \alpha$. (It is quite likely that y will remain $\frac{m}{s} - (1-\alpha)(V-2)$.)
(4) $x \leftarrow \frac{m}{s} - \alpha(V-1)$.
(5) If $x \leq \alpha$ then $x \leftarrow \alpha$. If $x \geq 1 - \alpha$ then $x \leftarrow 1 - \alpha$. (It is quite likely that x will remain $\frac{m}{s} - \alpha(V-1)$.)
(6) Output $((\alpha, x), (y, 1-\alpha))$.

The following picture captures the typical case:

$$(\; Vs_V \; V\text{-shs} \;)[\; 0 \;](\; (V-1)s_{V-1} \; (V-1)\text{-shs} \;)$$
$$\alpha \qquad\qquad\qquad x \quad y \qquad\qquad\qquad\qquad 1-\alpha$$

6.11. The Algorithm VHalf

The algorithm VHalf(m, s, α) tries to verify that $f(m, s) \leq \alpha$ can be proven by the Half method.

VHalf(m, s, α):

(1) Input(m, s, α).
(2) If $\alpha < \frac{1}{3}$ then output **Bad Input**.
(3) $(V, s_V, s_{V-1}) \leftarrow SV(m, s)$.
(4) If $\frac{m}{s} \times \frac{1}{V+1} > \alpha$ or $1 - \frac{m}{s} \times \frac{1}{V-2} > \alpha$ then output **DK** and stop. (If this happens then the V-conjecture did not hold. This has never occurred.)
(5) $((\alpha, x), (y, 1-\alpha)) \leftarrow$ FINDEND(m, s, α, V).
(6) If $x \leq \frac{1}{2}$ and $Vs_V > m$ then output **Yes** and stop.
 (In this case there are more than m shares that are $\leq \frac{1}{2}$.)
(7) If $y \geq \frac{1}{2}$ and $(V-1)s_{V-1} > m$ then output **Yes** and stop.
 (In this case there are more than m shares that are $\geq \frac{1}{2}$.)
(8) If you get here, then output **DK** and stop.

6.12. The Half Method When the Answer Is Not Known

From Sections 6.10.1 and 6.10.2 we can, given m, s, α, find V, s_V, s_{V-1}, x, y such that:

(1) There are s_{V-1} $(V-1)$-students, hence $(V-1)s_{V-1}$ $(V-1)$-shares.
(2) There are s_V V-students, hence Vs_V V-shares.
(3) The $(V-1)$-shares are all in $(y, 1-\alpha)$.
(4) The V-shares are all in (α, x).

In this section we turn this around. We *want* to have either $x = \frac{1}{2}$ or $y = \frac{1}{2}$ so that we can use the Half method. We will derive the value of α that makes that happen.

To make $x = \frac{1}{2}$ we need:

$$x = \frac{m}{s} - \alpha(V-1) = \frac{1}{2}$$

$$\alpha = \frac{\frac{m}{s} - \frac{1}{2}}{V-1}.$$

To make $y = \frac{1}{2}$ we need:

$$y = \frac{m}{s} - (1-\alpha)(V-2) = \frac{1}{2}$$

$$1 - \alpha = \frac{\frac{m}{s} - \frac{1}{2}}{V-2}$$

$$\alpha = 1 - \frac{\frac{m}{s} - \frac{1}{2}}{V-2}.$$

The two possible values of α above look promising. But there are two (minor) snags: (1) what if the $\frac{1}{2}$ is between x and y? (2) what if $\alpha < \frac{1}{3}$? It turns out these happen at the same time. If $\alpha < \frac{1}{3}$ then we can just take $\alpha = \frac{1}{3}$, as in Theorem 6.9. This is the only time that $\frac{1}{2}$ is between x and y.

Half(m, s):

(1) Input(m, s).
(2) If s divides m then output 1 and stop.
(3) $(V, s_V, s_{V-1}) \leftarrow SV(m, s)$.

(4) If $(V-1)s_{V-1} > Vs_V$ then $\alpha \leftarrow 1 - \frac{\frac{m}{s} - \frac{1}{2}}{V-2}$.
(5) If $\alpha < \frac{1}{3}$ then $\alpha \leftarrow \frac{1}{3}$.
(6) If VHalf$(m, s, \alpha) = $ **Yes** then output α.
(7) If $(V-1)s_{V-1} < Vs_V$ then $\alpha \leftarrow \frac{\frac{m}{s} - \frac{1}{2}}{V-1}$.
(8) If $\alpha < \frac{1}{3}$ then $\alpha \leftarrow \frac{1}{3}$.
(9) If VHalf$(m, s, \alpha) = $ **Yes** then output α.
(10) If none of the above hold then output 1 (this is a way of saying that the Half method does not produce a useful answer).

While the algorithm was originally assuming the V-conjecture, the VHalf step ensures that the algorithm works in any case. We have never seen a case where Half(m, s) produced α and VHalf(m, s, α) failed to verify.

Theorem 6.15. *For all* $m \geq s$, $f(m, s) \leq $ Half(m, s).

Exercise 6.16. Write programs for all of the algorithms in this section.

6.13. Program and Progress

Using the techniques presented so far we have the following attempt at an algorithm to find $f(m, s)$:

(1) Input(m, s).
(2) α is the min of FC(m, s), Half(m, s).
(3) Run FINDPROC(m, s, α). If it outputs a procedure P then output α, else output **DK**.

There are 3520 pairs (m, s) we are considering (see Chapter 3). There were 329 pairs that FC was unable to solve but Half did. Here are the full statistics so far. When we state that (say) for 329 cases $f(m, s) = $ Half(m, s), it is implicit that the prior techniques (in the case of Half its just FC) did not obtain the upper bound.

- For 2301 of them $f(m, s) = $ FC(m, s). This is $\sim 65.37\%$.
- For 329 of them $f(m, s) = $ Half(m, s). This is $\sim 9.35\%$.
- For 890 of them, the functions FC, Half, and FINDPROC did not suffice to find $f(m, s)$. This is $\sim 25.28\%$.

Chapter 7

A Formula for $f(m,5)$

7.1. Goals

We would like formulas for $f(m, 5)$ similar in spirit to the formulas for $f(m, 3)$ and $f(m, 4)$ in Theorems 4.9 and 4.11. For $f(m, 3)$ we do the following (in retrospect):

(1) For $m \equiv 1 \pmod{3}$ find a formula for FC$(m, 3)$.
(2) Prove that for all $m \equiv 1 \pmod 3$, $m \geq 4$, $f(m, 3) = $ FC$(m, 3)$.
(3) Do the same for $f(m, 3)$ where $m \equiv 2 \pmod 3$.
(4) The case of $f(m, 3)$ where $m \equiv 0 \pmod 3$ is easy.
(5) Use the above to get a formula for $f(m, 3)$ with three cases.

For $f(m, 4)$ we went through a similar process with $m \equiv 1, 3 \pmod 4$ being the hard cases and $m \equiv 0, 2 \pmod 4$ being the easy cases.

For $f(m, 5)$ we hope to do the following:

(1) For $m \equiv 1 \pmod 5$ find a formula for FC$(m, 5)$.
(2) Prove that for all $m \equiv 1 \pmod 5$, $f(m, 5) = $ FC$(m, 5)$.
(3) Do the same for $f(m, 5)$ where $m \equiv 2, 3, 4 \pmod 5$.
(4) The case of $m \equiv 0 \pmod 5$ is easy.
(5) Put together the cases to get a formula for $f(m, 5)$ with five cases.

That would be great! But there is one problem. Or maybe two. From Exercise 6.2 we saw that, for $m = 6, 7, 8, 9, 12, 13, 14$, $f(m, 5) = $ FC$(m, 5)$, but:

- $f(7, 5) = \frac{1}{3} = \text{FC}(7, 5)$. However, this is one of those rare cases where $\text{FC}(7, 5) = \frac{1}{3}$ instead of

$$\min\left\{\frac{m}{s} \times \frac{1}{\lceil \frac{2m}{s} \rceil},\ 1 - \frac{m}{s} \times \frac{1}{\lfloor \frac{2m}{s} \rfloor}\right\},$$

so $\frac{1}{3}$ might not fit into a nice pattern.
- $f(11, 5) = \frac{13}{30} < \text{FC}(11, 5)$. Hence this will clearly not fit the FC-pattern.

What do we do about these (m, s) which will likely upset the pattern? Let's turn this lemon into lemonade! We will come up with formulas and proofs that (usually) work; however, we will note what happens for $f(7, 5)$ and $f(11, 5)$. The proof of the formula should *not* work there, and it will be interesting to see why.

We use Notation 4.7 for when students get shares of different sizes. We give an example of its use.

Give 2 students

$$\left\{8k + 1 \Bigg\| \frac{30k + 4}{60k + 5} \Bigg\| 1 : \frac{30k + 2}{60k + 5} \Bigg\| 2k - 1 : 1\right\}$$

means that the 2 students get $8k+1$ shares of size $\frac{30k+4}{60k+5}$, 1 share of size $\frac{30k+2}{60k+5}$, and $2k - 1$ shares of size 1 (so $2k - 1$ muffins). This notation generalizes in the obvious way.

7.2. Finding the Formula for $f(m, 5)$

Exercise 7.1.

(1) For $1 \leq i \leq 4$ find a formula for $\text{FC}(5k+i, 5)$. Note that $f(5k+i, 5) \leq \text{FC}(5k+i, 5)$. For future parts of this section let $f_i(k) = \text{FC}(5k+i, 5)$.
(2) For $1 \leq i \leq 4$ prove that $f(5k + i, 5) = f_i(k)$ except for $f(7, 5)$ and $f(11, 5)$. Point out why your proof fails in those places. *Hint*: Get procedures for (say) $f(8, 5)$, $f(13, 5)$, $f(18, 5)$ and try to see similarities. *Hint*: The formulas are mod 5; however, some parts of the proof are mod 30. For example, the case of $f(5k + 1, 5)$ divides into subcases $f(30k+1, 5)$, $f(30k+6, 5)$, $f(30k+11, 5)$, $f(30k+16, 5)$.

$f(30k+21, 5)$, $f(30k+26, 5)$. *Advice*: Once this problem stops being fun, stop working on it and skip to the solution.

Solution to Exercise 7.1

The following theorem includes the formulas. We also include $f(5k, 5)$ for completeness.

Theorem 7.2. *If* $m = 5k + i$, *where* $0 \le i \le 4$, *then* $f(m, 5)$ *depends only on* k, i *via a formula, given below, with 2 exceptions (we will note the exceptions).*

Case 0: $m \equiv 0 \pmod 5$, *so* $m = 5k + 0$ *with* $k \ge 1$. *Then* $f(5k, 5) = 1$.

Case 1: $m \equiv 1 \pmod 5$, *so* $m = 5k + 1$ *with* $k \ge 1$, $k \ne 2$. *Then* $f(5k + 1, 5) = \frac{5k+1}{10k+5}$. (*Exception*: $f(11, 5) = \frac{13}{30} < \frac{11}{25}$.)

Case 2: $m \equiv 2 \pmod 5$, *so* $m = 5k + 2$ *with* $k \ge 2$. *Then* $f(5k+2, 5) = \frac{5k-2}{10k}$. (*Exception*: $f(7, 5) = \frac{1}{3} > \frac{3}{10}$.)

Case 3: $m \equiv 3 \pmod 5$, *so* $m = 5k + 3$ *with* $k \ge 1$. *Then* $f(5k+3, 5) = \frac{5k+3}{10k+10}$.

Case 4: $m \equiv 4 \pmod 5$, *so* $m = 5k + 4$ *with* $k \ge 1$. *Then* $f(5k+4, 5) = \frac{5k+1}{10k+5}$.

Proof. All of the upper bounds are by the Floor–Ceiling Theorem. We give the lower bounds by giving procedures. We allow some muffins to be uncut in the procedures. The mantra that all muffins are cut into exactly 2 pieces is needed for upper bounds, not lower bounds.

Case 1: $m = 5k + 1$ with $k \ge 1$, $k \ne 2$. Then $f(5k + 1, 5) = \frac{5k+1}{10k+5}$. There are cases for mod 30.

- $m = 30k + 1$ with $k \ge 1$, $f(30k + 1, 5) = \frac{30k+1}{60k+5}$.

  (1) Divide $24k + 2$ muffins $\left\{\frac{30k+1}{60k+5}, \frac{30k+4}{60k+5}\right\}$.
  (2) Divide 2 muffins $\left\{\frac{30k+2}{60k+5}, \frac{30k+3}{60k+5}\right\}$.
  (3) Leave $6k - 3$ muffins uncut.
  (4) Give 2 students $\left\{12k + 1 : \frac{30k+1}{60k+5}\right\}$.

(5) Give 2 students $\{8k+1: \frac{30k+4}{60k+5} \mid\mid 1: \frac{30k+2}{60k+5} \mid\mid 2k-1:1\}$.

(6) Give 1 student $\{8k: \frac{30k+4}{60k+5} \mid\mid 2: \frac{30k+3}{60k+5} \mid\mid 2k-1:1\}$.

- $m = 30k+6$ with $k \geq 0$, $f(30k+6, 5) = \frac{30k+6}{60k+15}$.

  (1) Divide $24k+6$ muffins $\{\frac{30k+6}{60k+15}, \frac{30k+9}{60k+15}\}$.

  (2) Leave $6k$ muffins uncut.

  (3) Give 2 students $\{12k+3: \frac{30k+6}{60k+15}\}$.

  (4) Give 3 students $\{8k+2: \frac{30k+9}{60k+15} \mid\mid 2k:1\}$.

- $m = 30k+11$ with $k \geq 1$, $f(30k+11, 5) = \frac{30k+11}{60k+25}$.

  (1) Divide $24k+10$ muffins $\{\frac{30k+11}{60k+25}, \frac{30k+14}{60k+25}\}$.

  (2) Divide 2 muffins $\{\frac{30k+12}{60k+25}, \frac{30k+13}{60k+25}\}$.

  (3) Leave $6k-1$ muffins uncut.

  (4) Give 2 students $\{12k+5: \frac{30k+11}{60k+25}\}$.

  (5) Give 2 students $\{8k+3: \frac{30k+14}{60k+25} \mid\mid 1: \frac{30k+13}{60k+25} \mid\mid 2k:1\}$.

  (6) Give 1 student $\{8k+4: \frac{30k+14}{60k+25} \mid\mid 2: \frac{30k+12}{60k+25} \mid\mid 2k-1:1\}$.

Note 7.3. When looking at $f(30k+11, 5)$, the $k=0$ case should *not work* since we know $f(11, 5) = \frac{13}{30} < \frac{11}{25}$. Hence the above prove *should fail* when $k=0$. Indeed it does. Note the last line, when $k=0$, would involve giving someone -1 muffins. Nobody deserves -1 muffins!

- $m = 30k+16$ with $k \geq 0$, $f(30k+16, 5) = \frac{30k+16}{60k+35}$.

  (1) Divide $24k+14$ muffins $\{\frac{30k+16}{60k+35}, \frac{30k+19}{60k+35}\}$.

  (2) Divide 2 muffins $\{\frac{30k+17}{60k+35}, \frac{30k+18}{60k+35}\}$.

  (3) Leave $6k$ muffins uncut.

  (4) Give 2 students $\{12k+7: \frac{30k+16}{60k+35}\}$.

  (5) Give 2 students $\{8k+5: \frac{30k+19}{60k+35} \mid\mid 1: \frac{30k+17}{60k+35} \mid\mid 2k:1\}$.

  (6) Give 1 student $\{8k+4: \frac{30k+19}{60k+35} \mid\mid 2: \frac{30k+18}{60k+35} \mid\mid 2k:1\}$.

- $m = 30k+21$ with $k \geq 0$, $f(30k+21, 5) = \frac{10k+7}{20k+15}$.

  (1) Divide $24k+18$ muffins $\{\frac{10k+7}{20k+15}, \frac{10k+8}{20k+15}\}$.

  (2) Leave $6k+3$ muffins uncut.

(3) Give 2 students $\{12k+9 : \frac{10k+7}{20k+15}\}$.

(4) Give 3 students $\{8k+6 : \frac{10k+8}{20k+15} \,\|\, 2k+1:1\}$.

- $m = 30k+26$ with $k \geq 0$, $f(30k+26, 5) = \frac{30k+26}{60k+55}$.

  (1) Divide $24k+22$ muffins $\{\frac{30k+26}{60k+55}, \frac{30k+29}{60k+55}\}$.

  (2) Divide 2 muffins $\{\frac{30k+27}{60k+55}, \frac{30k+28}{60k+55}\}$.

  (3) Leave $6k+2$ uncut.

  (4) Give 2 students $\{12k+11 : \frac{30k+26}{60k+55}\}$.

  (5) Give 2 students $\{8k+7 : \frac{30k+29}{60k+55} \,\|\, 1 : \frac{30k+28}{60k+55} \,\|\, 2k+1:1\}$.

  (6) Give 1 student $\{8k+8 : \frac{30k+29}{60k+55} \,\|\, 2 : \frac{30k+27}{60k+55} \,\|\, 2k:1\}$.

Case 2: $m = 5k+2$ with $k \geq 2$. Then $f(5k+2, 5) = \frac{5k-2}{10k}$. There are cases for mod 10.

- $m = 10k+2$ with $k \geq 1$, $f(10k+2, 5) = \frac{10k-2}{20k}$.

  (1) Divide $4k$ muffins $\{\frac{10k-2}{20k}, \frac{10k+2}{20k}\}$.

  (2) Divide $6k+2$ muffins $\{\frac{1}{2}, \frac{1}{2}\}$.

  (3) Give 1 student $\{4k : \frac{10k+2}{20k}\}$.

  (4) Give 4 students $\{k : \frac{10k-2}{20k} \,\|\, 3k+1 : \frac{1}{2}\}$.

- $m = 10k+7$ with $k \geq 1$, $f(10k+7, 5) = \frac{10k+3}{20k+10}$.

  (1) Divide $4k+2$ muffins $\{\frac{10k+3}{20k+10}, \frac{10k+7}{20k+10}\}$.

  (2) Divide $4k+2$ muffins $\{\frac{10k+4}{20k+10}, \frac{10k+6}{20k+10}\}$.

  (3) Divide $2k+3$ muffins $\{\frac{1}{2}, \frac{1}{2}\}$.

  (4) Give 1 student $\{4k+2 : \frac{10k+7}{20k+10}\}$.

  (5) Give 2 students $\{2k+1 : \frac{10k+3}{20k+10} \,\|\, 2k+1 : \frac{10k+6}{20k+10} \,\|\, 1 : \frac{1}{2}\}$.

  (6) Give 2 students $\{2k+1 : \frac{10k+4}{20k+10} \,\|\, 2k+2 : \frac{1}{2}\}$.

Note 7.4. When looking at $f(10k+7)$, the $k=0$ case *does work*. Hence, by the above, we can deduce $f(7, 5) \geq \frac{3}{10}$. This is true, but not optimal. Recall that Solution 6.2 showed $f(7, 5) = \frac{1}{3}$.

Case 3: $m = 5k+3$ with $k \geq 1$. Then $f(5k+3, 5) = \frac{5k+3}{10k+10}$. There are three cases.

- $m = 10k + 3$ with $k \geq 1$, $f(10k+3, 5) = \frac{10k+3}{20k+10}$.
    (1) Divide $4k + 2$ muffins $\left\{\frac{10k+3}{20k+10}, \frac{10k+7}{20k+10}\right\}$.
    (2) Divide 3 muffins $\left\{\frac{10k+4}{20k+10}, \frac{10k+6}{20k+10}\right\}$.
    (3) Divide $6k - 2$ muffins $\left\{\frac{1}{2}, \frac{1}{2}\right\}$.
    (4) Give 1 student $\left\{4k + 2 : \frac{10k+3}{20k+10}\right\}$.
    (5) Give 3 students $\left\{\frac{10k+4}{20k+10} \,\middle\|\, k+1 : \frac{10k+7}{20k+10} \,\middle\|\, 3k-1 : \frac{1}{2}\right\}$.
    (6) Give 1 student $\left\{3 : \frac{10k+6}{20k+10} \,\middle\|\, k-1 : \frac{10k+7}{20k+10} \,\middle\|\, 3k-1 : \frac{1}{2}\right\}$.
- $m = 20k + 8$ with $k \geq 1$, $f(20k+8, 5) = \frac{5k+2}{10k+5}$.
    (1) Divide $20k + 8$ muffins $\left\{\frac{5k+2}{10k+5}, \frac{5k+3}{10k+5}\right\}$.
    (2) Give 1 student $\left\{8k + 4 : \frac{5k+2}{10k+5}\right\}$.
    (3) Give 4 students $\left\{3k + 1 : \frac{5k+2}{10k+5} \,\middle\|\, 5k+2 : \frac{5k+3}{10k+5}\right\}$.
- $m = 20k + 18$ with $k \geq 1$, $f(20k+18, 5) = \frac{10k+9}{20k+20}$.
    (1) Divide $8k + 8$ muffins $\left\{\frac{10k+9}{20k+20}, \frac{10k+11}{20k+20}\right\}$.
    (2) Divide $12k + 10$ muffins $\left\{\frac{10k+10}{20k+20}, \frac{10k+10}{20k+20}\right\}$.
    (3) Give 1 student $\left\{8k + 8 : \frac{10k+9}{20k+20}\right\}$.
    (4) Give 4 students $\left\{6k + 5 : \frac{10k+10}{20k+20}, 2k+2 : \frac{20k+11}{20k+20}\right\}$.

Case 4: $m = 5k + 4$ with $k \geq 1$. Then $f(5k+4, 5) = \frac{5k+1}{10k+5}$. There are cases for mod 30.

- $m = 30k + 4$ with $k \geq 1$, $f(30k+4, 5) = \frac{30k+1}{60k+5}$.
    (1) Divide $24k + 2$ muffins $\left\{\frac{30k+1}{60k+5}, \frac{30k+4}{60k+5}\right\}$.
    (2) Divide 2 muffins $\left\{\frac{30k+2}{60k+5}, \frac{30k+3}{60k+5}\right\}$.
    (3) Leave $6k$ muffins uncut.
    (4) Give 2 students $\left\{12k + 1 : \frac{30k+4}{60k+5}\right\}$.
    (5) Give 2 students $\left\{8k + 1 : \frac{30k+1}{60k+5} \,\middle\|\, 1 : \frac{30k+3}{60k+5} \,\middle\|\, 2k : 1\right\}$.
    (6) Give 1 student $\left\{8k : \frac{30k+1}{60k+5} \,\middle\|\, 2 : \frac{30k+2}{60k+5} \,\middle\|\, 2k : 1\right\}$.
- $m = 30k + 9$ with $k \geq 0$, $f(30k+9, 5) = \frac{10k+2}{20k+5}$.
    (1) Divide $24k + 6$ muffins $\left\{\frac{10k+2}{20k+5}, \frac{10k+3}{20k+5}\right\}$.
    (2) Leave $6k + 3$ muffins uncut.

A Formula for $f(m, 5)$ 73

   (3) Give 2 students $\{12k + 3 : \frac{10k+3}{20k+5}\}$.
   (4) Give 3 students $\{8k + 2 : \frac{10k+2}{20k+5} \,\|\, 2k + 1 : 1\}$.

- $m = 30k + 14$ with $k \geq 0$, $f(30k + 14, 5) = \frac{30k+11}{60k+25}$.

   (1) Divide $24k + 10$ muffins $\{\frac{30k+11}{60k+25}, \frac{30k+14}{60k+25}\}$.
   (2) Divide 2 muffins $\{\frac{30k+12}{60k+25}, \frac{30k+13}{60k+25}\}$.
   (3) Leave $6k + 2$ muffins uncut.
   (4) Give 2 students $\{12k + 5 : \frac{30k+14}{60k+25}\}$.
   (5) Give 2 students $\{8k + 3 : \frac{30k+11}{60k+25} \,\|\, 1 : \frac{30k+12}{60k+25} \,\|\, 2k + 1 : 1\}$.
   (6) Give 1 student $\{8k + 4 : \frac{30k+11}{60k+25} \,\|\, 2 : \frac{30k+13}{60k+25} \,\|\, 2k : 1\}$.

- $m = 30k + 19$ with $k \geq 0$, $f(30k + 19, 5) = \frac{30k+16}{60k+35}$.

   (1) Divide $24k + 14$ muffins $\{\frac{30k+16}{60k+35}, \frac{30k+19}{60k+35}\}$.
   (2) Divide 2 muffins $\{\frac{30k+17}{60k+35}, \frac{30k+18}{60k+35}\}$.
   (3) Leave $6k + 3$ muffins uncut.
   (4) Give 2 students $\{12k + 7 : \frac{30k+19}{60k+35}\}$.
   (5) Give 2 students $\{8k + 5 : \frac{30k+16}{60k+35} \,\|\, 1 : \frac{30k+18}{60k+35} \,\|\, 2k + 1 : 1\}$.
   (6) Give 1 student $\{8k + 4 : \frac{30k+16}{60k+35} \,\|\, 2 : \frac{30k+17}{60k+35} \,\|\, 2k + 1 : 1\}$.

- $m = 30k + 24$ with $k \geq 0$, $f(30k + 24, 5) = \frac{10k+7}{20k+15}$.

   (1) Divide $24k + 18$ muffins $\{\frac{10k+7}{20k+15}, \frac{10k+8}{20k+15}\}$.
   (2) Divide $6k + 6$ muffins uncut.
   (3) Give 2 students $\{12k + 9 : \frac{10k+8}{20k+15}\}$.
   (4) Give 3 students $\{8k + 6 : \frac{10k+7}{20k+15} \,\|\, 2k + 2 : 1\}$,

- $m = 30k + 29$ with $k \geq 0$, $f(30k + 29, 5) = \frac{30k+26}{60k+55}$.

   (1) Divide $24k + 22$ muffins $\{\frac{30k+26}{60k+55}, \frac{30k+29}{60k+55}\}$.
   (2) Divide 2 students $\{\frac{30k+27}{60k+55}, \frac{30k+28}{60k+55}\}$.
   (3) Leave $6k + 5$ muffins uncut.
   (4) Give 2 students $\{12k + 11 : \frac{30k+29}{60k+55}\}$.
   (5) Give 2 students $\{8k + 7 : \frac{30k+26}{60k+55} \,\|\, 1 : \frac{30k+27}{60k+55} \,\|\, 2k + 2 : 1\}$.
   (6) Give 1 student $\{8k + 8 : \frac{30k+26}{60k+55} \,\|\, 2 : \frac{30k+28}{60k+55} \,\|\, 2k + 1 : 1\}$.   □

7.3. Formulas for Six, Seven, Eight, Nine

Ken Tan, a high school student interested in muffins (both eating them and working out formulas for them), spent the summer of 2018 working out the formulas for $f(m, 6)$, $f(m, 7)$, $f(m, 8)$, and $f(m, 9)$. He did it by hand. The formulas are too long to be in this book; however, they will be at the MUFFIN website.

7.4. Why Stop Here?

In the summer of 2019 I had another student, Alex Kwang, working on the case of $f(10k + i, 10)$. We know all of the answers from FC and our data. Even so, it seemed like proving the answer (that is, getting procedures) would be mod... a very large number. So it looks like $f(9k + i, 9)$ is a good place to stop.

Alex then worked on $f(m, s)$ where $m - s$ is a constant. We discuss upper bounds for such problems in Chapter 10. Alex obtained many lower bounds. His paper is on the MUFFIN website.

Chapter 8

The Interval Method

8.1. Recap and Goals for This Chapter

From Chapters 4 and 6, we know that, for $1 \leq s \leq m$,

$$f(m, s) \leq \min\{\text{FC}(m, s), \text{Half}(m, s)\}.$$

Just for now we will refer to the "min" of those quantities as $\text{minf}(m, s)$. Is it the case that, for all $m \geq s$, $f(m, s) = \text{minf}(m, s)$? No. The counterexample with the smallest s is $f(10, 9)$.

We show the following:

$$f(10, 9) \leq \frac{1}{3} < \text{minf}(10, 9),$$

$$f(24, 11) \leq \frac{19}{44} < \text{minf}(24, 11),$$

$$f(16, 13) \leq \frac{14}{39} < \text{minf}(16, 13).$$

The proof of the upper bounds uses a technique, which we call *The Interval Method* (*INT method for short*). We develop an algorithm $\text{INT}(m, s)$ which on input m, s gives an α such that $f(m, s) \leq \alpha$.

Exercise 8.1. For $(m, s) = (10, 9), (24, 11), (16, 13)$:

(1) Compute $\alpha = \text{minf}(m, s)$.
(2) Compute $\text{FINDPROC}(m, s, \alpha)$. (You should get a **DK** which means that $\text{minf}(m, s)$ is unlikely to be $f(m, s)$.)

8.2. Ten Muffins, Nine Students

Theorem 8.2. $f(10, 9) = \frac{1}{3}$.

Proof. We leave the proof that $f(10, 9) \geq \frac{1}{3}$ to the reader. Alternatively, the reader can run FINDPROC(10, 9, $\frac{1}{3}$) or use Theorem 4.5.

We now show $f(10, 9) \leq \frac{1}{3}$. Assume, by way of contradiction, that there is a (10, 9)-procedure with smallest piece $> \frac{1}{3}$. By Theorem 2.6, every muffin is cut into exactly 2 pieces. Hence there are 20 pieces.

We leave it to the reader to show that there are seven 2-students, two 3-students, fourteen 2-shares, six 3-shares, and that the following picture captures what we know:

$$(\text{ 6 3-shs })(\text{ 14 2-shs })$$
$$\frac{3}{9} \qquad\qquad \frac{4}{9} \qquad\qquad \frac{6}{9}$$

Note that the Half method won't work here since $\frac{1}{2} = \frac{4.5}{9}$ which is inside the interval with more shares.

There is no share of size $\frac{4}{9}$. Can we use that? The following is important and we will use this kind of reasoning often:

Since there is no share of size $\frac{4}{9}$, there is no share of size $\frac{5}{9}$, by buddying.

We will justify the following picture after presenting it.

$$(\text{ 6 3-shs })(\text{ 8 2-shs })(\text{ 6 2-shs })$$
$$\frac{3}{9} \qquad\quad \frac{4}{9} \qquad\quad \frac{5}{9} \qquad\quad \frac{6}{9}$$

- $(\frac{3}{9}, \frac{4}{9})$ contains all six 3-shares.
- Since $(\frac{3}{9}, \frac{4}{9})$ and $(\frac{5}{9}, \frac{6}{9})$ are buddies, and $(\frac{3}{9}, \frac{4}{9})$ has 6 shares, $(\frac{5}{9}, \frac{6}{9})$ contains 6 shares. They are 2-shares.
- There are fourteen 2-shares. Six of them are in $(\frac{5}{9}, \frac{6}{9})$, hence the remaining eight of them are in $(\frac{4}{9}, \frac{5}{9})$.

The 2-shares are in 2 open intervals: $(\frac{4}{9}, \frac{5}{9})$ and $(\frac{5}{9}, \frac{6}{9})$. We examine how many shares a 2-student can have from each interval.

If a 2-student has both shares from $(\frac{4}{9}, \frac{5}{9})$, then she has

$$< 2 \times \frac{5}{9} = \frac{10}{9}.$$

Hence this cannot occur. We now know that each of the seven 2-student takes at least one 2-share from $(\frac{5}{9}, \frac{6}{9})$. But $(\frac{5}{9}, \frac{6}{9})$ only has six shares. This is a contradiction. □

8.3. Twenty Four Muffins, Eleven Students

We show $f(24, 11) = \frac{19}{44}$.

Theorem 8.3. $f(24, 11) = \frac{19}{44}$.

Proof. We leave the proof that $f(24, 11) \geq \frac{19}{44}$ to the reader. Alternatively, the reader can run FINDPROC(24, 11, $\frac{19}{44}$).

We derive the upper bound pretending we do not know it. Let α be the upper bound. We assume $\alpha > \frac{1}{3}$. By Theorem 2.6 every muffin is cut into 2 pieces; hence there are 48 pieces. Assuming the V-conjecture we guess everyone is either a 4-student or a 5-student. We leave it to the reader to show that there are seven 4-students, four 5-students, twenty-eight 4-shares and twenty 5-shares.

By buddying, since the smallest share is $> \alpha$, the largest share is $< 1-\alpha$. If the smallest 4-share is of size y, then

$$3(1 - \alpha) + y > \frac{24}{11}$$

so $y > 3\alpha - \frac{9}{11}$. Hence $3\alpha - \frac{9}{11}$ is the left endpoint of the 4-share interval. If the largest 5-share is of size x, then

$$4\alpha + x < \frac{24}{11}$$

so $x < \frac{24}{11} - 4\alpha$.

The following picture captures what we know:

$$(\quad 20\ \text{5-shs}\quad)[\quad 0\quad](\quad 28\ \text{4-shs}\quad)$$
$$\alpha \qquad \tfrac{24}{11} - 4\alpha \qquad 3\alpha - \tfrac{9}{11} \qquad 1 - \alpha$$

Since there are no shares in $[\tfrac{24}{11} - 4\alpha, 3\alpha - \tfrac{9}{11}]$, by buddying, there are no shares in $[\tfrac{20}{11} - 3\alpha, 4\alpha - \tfrac{13}{11}]$. Since there are more shares on the right, we will guess that $3\alpha - \tfrac{9}{11} \leq \tfrac{1}{2}$ so this new empty interval is within the 4-shares. Hence the following picture represents what we know:

$$(\quad 20\ \text{5-shs}\quad)[\quad 0\quad](\quad 8\ \text{4-shs}\quad)[\quad 0\quad](\quad 20\ \text{4-shs}\quad)$$
$$\alpha \qquad \tfrac{24}{11} - 4\alpha \quad 3\alpha - \tfrac{9}{11} \qquad \tfrac{20}{11} - 3\alpha \quad 4\alpha - \tfrac{13}{11} \qquad 1 - \alpha$$

We call the shares in the first interval of 4-shares *small* and the shares in the second interval of 4-shares *large*. Note that Alice, a 4-student, has either (1) 4 small shares and 0 large shares, or ... (5) 0 small shares and 4 large shares. We want that Alice needs at least 3 large shares. Why? Because if Alice (and hence any 4-student) needs at least 3 large shares, and there are seven 4-students, there will have to be 21 large shares. Since there are 20, this will cause a contradiction.

We find α such that Alice has to have ≥ 3 large shares. If she had ≤ 2 large shares, then she has

$$< 2 \times \left(\frac{20}{11} - 3\alpha \right) + 2(1 - \alpha) \leq \frac{24}{11}$$

$$\alpha \geq \frac{19}{44}.$$

Hence, we have derived the bound $f(24, 11) \leq \tfrac{19}{44}$. $\square$

8.4. $f(16, 13) \leq \tfrac{14}{39}$

In Theorem 8.2 we proved $f(10, 9) = \tfrac{1}{3}$. The procedure used 2-shares and 3-shares, and the proof concentrated on the 2-shares. In Theorem 8.3 we proved $f(24, 11) = \tfrac{19}{44}$. The procedure used 4-shares and 5-shares, and the proof concentrated on the 4-shares. Do we always use the V-shares? No, as the following proof demonstrates.

Theorem 8.4. $f(16, 13) = \tfrac{14}{39}$.

The Interval Method 79

Proof. We leave the proof that $f(16, 13) \geq \frac{14}{39}$ to the reader. Alternatively, the reader can run FINDPROC$(16, 13, \frac{14}{39})$.

Let α be the upper bound. We assume $\alpha > \frac{1}{3}$. By Theorem 2.6 every muffin is cut into two pieces; hence there are 32 pieces. We assume the V-conjecture hence we guess everyone is either a 2-student or a 3-student. We leave it to the reader to show that there are seven 2-students, six 3-students, fourteen 2-shares, and eighteen 3-shares.

By buddying, since the smallest share is $> \alpha$, the largest share is $< 1-\alpha$. If the smallest 2-share is of size y then $(1 - \alpha) + y > \frac{16}{13}$, so $y > \frac{3}{13} + \alpha$. Hence $\frac{3}{13} + \alpha$ is the left endpoint of the 2-share interval. If the largest 3-share is of size x then $2\alpha + x < \frac{16}{13}$, so $x < \frac{16}{13} - 2\alpha$.

The following picture captures what we know:

$$(\quad 18 \text{ 3-shs} \quad)[\quad\quad 0 \quad](\quad 14 \text{ 2-shs} \quad)$$
$$\alpha \quad\quad\quad \tfrac{48}{39}-2\alpha \quad\quad \alpha+\tfrac{9}{39} \quad\quad\quad 1-\alpha$$

Since there are no shares in $[\frac{48}{39} - 2\alpha, \alpha + \frac{9}{39}]$, by buddying, there are no shares in $[\frac{30}{39} - \alpha, 2\alpha - \frac{9}{39}]$.

Is the gap $[\frac{30}{39} - \alpha, 2\alpha - \frac{9}{39}]$ going to be within the 2-shares or within the 3-shares? There are two cases.

Case 1: The gap is in the 2-shares. Then

$$\alpha + \frac{9}{39} \leq \frac{30}{39} - \alpha$$

$$2\alpha \leq \frac{21}{39}$$

$$\alpha \leq \frac{10.5}{39} < \frac{1}{3} < \frac{14}{39}.$$

Case 2: The gap is within the 3-shares.

The following picture captures what we know:

$$(\quad 14 \text{ 3-shs} \quad)[\quad 0 \quad](\quad 4 \text{ 3-shs} \quad)[\quad 0 \quad](\quad 14 \text{ 2-shs} \quad)$$
$$\alpha \quad\quad \tfrac{30}{39}-\alpha \quad 2\alpha-\tfrac{9}{39} \quad\quad \tfrac{48}{39}-2\alpha \quad \alpha+\tfrac{9}{39} \quad\quad\quad 1-\alpha$$

We (as usual) call the shares in the first interval of 3-shares *small* and the shares in the second interval of 3-shares *large*. Note that Alice, a 3-student, has either (1) 3 small shares and 0 large shares, or ... (4) 0 small shares and 3 large shares. We want that Alice needs at least 1 large share. Why? Because if Alice (and hence any 3-student) needs at least 1 large share, and there are six 3-students, there will be at least 6 large shares. Since there are only 4, this will cause a contradiction.

We find α such that Alice has to have ≥ 1 large share. If she had ≤ 0 large shares, then she has

$$3 \times \left(\frac{30}{39} - \alpha\right) \leq \frac{48}{39}$$

$$\alpha \geq \frac{14}{39}.$$

We have derived the upper bound $f(16, 13) \leq \frac{14}{39}$. □

8.5. Where Will the New Gap Be?

In the proofs of $f(10, 9) \leq \frac{1}{3}$, $f(24, 11) \leq \frac{19}{44}$, and $f(16, 13) \leq \frac{14}{39}$, we do the following:

(1) Let $V = \lceil \frac{2m}{s} \rceil$. We assume the V-conjecture; hence everyone is either a V-student or $(V-1)$-student. Find how many students there are of each type and hence how many shares there are of each type.
(2) Find the intervals that contain the $(V-1)$-shares and the V-shares. If they are not disjoint then quit and output **DK**. (In this case it is likely that the FC method or the Half method can be used to prove the upper bound.)
(3) By buddying, we found another gap. We call this *the new gap*.
(4) The new gap was either within the $(V-1)$-shares or the V-shares. Whichever interval the gap was in, there are now two intervals.
(5) We show that one of the intervals needs to have more shares than it does, hence getting a contradiction.

We need to know which set of shares the new gap will be within. Note the following:

(1) In the proof of $f(10, 9) \leq \frac{1}{3}$ the new gap was in the 2-shares. There were more 2-shares than 3-shares. (The new gap was only one point, but this is not important here.)
(2) In the proof of $f(24, 11) \leq \frac{19}{44}$ the new gap was in the 4-shares. There were more 4-shares than 5-shares.
(3) In the proof of $f(16, 13) \leq \frac{14}{39}$ the new gap was in the 3-shares. There were more 3-shares than 2-shares.

We have also observed the following empirically:

The only cases where there are equal numbers of $(V-1)$-shares and V-shares is when there exists a k such that $m = \frac{k^2+3k+8}{2}$ and $s = 2k+1$. In these cases $f(m, s) = \mathrm{FC}(m, s) = \frac{k+1}{2k+3}$.

Conjecture 8.5 (The VV-conjecture). *Assume $m > s$ and $f(m, s) > \frac{1}{3}$. There is an optimal (m, s)-procedure such that the following conditions hold:*

(1) *There is a V such that everyone is either a $(V-1)$-student or a V-student. (This is the V-conjecture.)*
(2) *Assume that the V-interval and the $(V-1)$-interval are disjoint.*
    (a) *If there are more V-shares than $(V-1)$-shares, then the new gap will be within the V-shares.*
    (b) *If there are more $(V-1)$-shares than V-shares, then the new gap will be within the $(V-1)$-shares.*
    (c) *If there are the same number of V-shares as $(V-1)$-shares, then $f(m, s) = \mathrm{FC}(m, s)$, and hence, we do not need to use the INT method.*

8.6. Exercises on Deriving or Verifying

Exercise 8.6. Prove each of the following statements of the form $f(m, s) \leq \alpha$ using the techniques of this chapter. Alternatively, try to not look at the upper bounds and instead derive it.

(1) $f(24, 11) \leq \frac{19}{44}$.
(2) $f(59, 14) \leq \frac{131}{280}$.

(3) $f(17, 15) \le \frac{7}{20}$.
(4) $f(19, 17) \le \frac{1}{3}$.
(5) $f(21, 17) \le \frac{6}{17}$.
(6) $f(21, 19) \le \frac{13}{38}$.
(7) $f(61, 19) \le \frac{313}{684}$.
(8) $f(33, 20) \le \frac{41}{100}$.

Note 8.7. All of the upper bounds in Exercise 8.6 are actually lower bounds. Hence we know, for example, that $f(33, 20) = \frac{41}{100}$.

8.7. Algorithms for INT: You Write It

Exercise 8.8.

(1) Let VINT take as input m, s, α and try to verify $f(m, s) \le \alpha$ via the INT method. Write an algorithm for VINT. It should output (1) **Yes** if $f(m, s) \le \alpha$ can be proven with the INT method, and (2) **DK** if $f(m, s) \le \alpha$ cannot be established with the INT method.
(2) Let INT take as input (m, s), derive a good candidate α for the INT method, and then apply VINT. Hence it will either (1) output an α and verify it and output **alpha** (2) output 1 if VINT(m, s, α) is **DK**.

8.8. The INT Theorem

Definition 8.9. INT(m, s) is defined as follows:

(1) If s divides m then output 1.
(2) If not then run the program INT(m, s). If it outputs that the α it produces does not work, then output 1. If it outputs that the α it produces does work, then output α.

Theorem 8.10. *If $m \le s$ then $f(m, s) \le$ INT(m, s).*

8.9. Program and Progress

Using the techniques presented so far we have the following attempt at an algorithm to find $f(m, s)$:

(1) Input(m, s).
(2) α is the min of
$$\{FC(m, s), \text{Half}(m, s), \text{INT}(m, s)\}.$$
(3) Run FINDPROC(m, s, α). If it outputs a procedure P then output α, else output **DK**.

There are 3520 pairs (m, s) we are considering (see Chapter 3). There were 186 pairs that neither FC nor Half were able to solve, but INT was. Here are the full statistics so far. When we state that (say) for 329 cases $f(m, s) = \text{Half}(m, s)$, it is implicit that the prior techniques (in the case of Half its just FC) did not obtain the upper bound.

- For 2301 of them, $f(m, s) = FC(m, s)$. This is $\sim 65.37\%$.
- For 329 of them, $f(m, s) = \text{Half}(m, s)$. This is $\sim 9.35\%$.
- For 186 of them, $f(m, s) = \text{INT}(m, s)$. This is $\sim 5.28\%$.
- For 704 of them, FC, Half, INT did not suffice to find $f(m, s)$. This is $\sim 20.00\%$.

Chapter 9

The Midpoint Method

9.1. Recap and Goals for This Chapter

From Chapters 4, 6, and 8 we know that, for $m \geq s$,

$$f(m, s) \leq \min\{\text{FC}(m, s), \text{Half}(m, s), \text{INT}(m, s)\}.$$

Just for now we will refer to the min of those quantities as $\text{minf}(m, s)$. Is it the case that, for all $m \geq s$, $f(m, s) = \text{minf}(m, s)$? No. The counterexample with the smallest s is $f(13, 12)$; however it is more illustrative to show the following:

$$f(23, 13) \leq \frac{53}{130} < \text{minf}(23, 13).$$

This proof uses a new technique, which we call *The Midpoint Method* (*MID method for short*). We develop it fully and create a program $\text{MID}(m, s)$ which, given m, s, outputs α such that $f(m, s) \leq \alpha$. MID is an extension of INT.

Exercise 9.1. For $(m, s) = (13, 12), (14, 13), (23, 13), (23, 14)$:

(1) Compute $\alpha = \text{minf}(m, s)$.
(2) Compute $\text{FINDPROC}(m, s, \alpha)$. (You should get a **DK** which means that $\text{minf}(m, s)$ is unlikely to be $f(m, s)$.)

9.2. Twenty-Three Muffins, Thirteen Students

We use the following notation in the rest of the book.

Notation 9.2. If (a, b) is an interval then $|(a, b)|$ is the number of shares in that interval. Note that this is not standard.

Theorem 9.3. $f(23, 13) = \frac{53}{130}$.

Proof. We leave the proof that $f(23, 13) \geq \frac{53}{130}$ to the reader. Alternatively, the reader can run FINDPROC(23, 13, $\frac{53}{130}$).

We begin the proof that $f(23, 13) \leq \frac{53}{130}$ as if we are using the INT method.

Assume, by way of contradiction, that there is a (23, 13)-procedure with smallest piece $> \frac{53}{130}$. By Theorem 2.6, every muffin is cut into exactly 2 pieces. Everyone gets $\frac{23}{13} = \frac{230}{130}$. there are six 3-students, seven 4-students, eighteen 3-shares, and twenty-eight 4-shares, and that the following picture captures what we know:

$$(\text{ 28 4-shs })[\ 0\ ](\ 18\ 3\text{-shs }\ )$$
$$\frac{53}{130} \qquad\qquad \frac{71}{130}\quad \frac{76}{130} \qquad\qquad \frac{77}{130}$$

Since $[\frac{71}{130}, \frac{76}{130}]$ is empty, by buddying, $[\frac{54}{130}, \frac{59}{130}]$ is empty. Hence the following picture captures what we know:

$$(\text{ 18 4-shs })[\ 0\ ](\ 10\ 4\text{-shs })[\ 0\ ](\ 18\ 3\text{-shs }\ )$$
$$\frac{53}{130} \qquad \frac{54}{130}\quad \frac{59}{130} \qquad\qquad \frac{71}{130}\quad \frac{76}{130} \qquad\qquad \frac{77}{130}$$

We call the first interval of 4-shares *small shares* and the second interval of 4-shares *large shares*. Let Alice be a 4-student.

- If Alice has 4 small shares and 0 large shares then she has $< 4 \times \frac{54}{130} = \frac{216}{130} < \frac{230}{130}$. So this is impossible.
- If Alice has 3 small shares and 1 large share then she has $< 3 \times \frac{54}{130} + \frac{71}{130} = \frac{233}{130}$. So she does get enough. Does she get too much? She gets $\geq 3 \times \frac{53}{130} + \frac{59}{130} = \frac{218}{130}$. This is not too much. This kind of student is possible.
- If Alice has 2 small shares and 2 large shares then by the last case she has enough. Does she have too much? She has $> 2 \times \frac{53}{130} + 2 \times \frac{59}{130} = \frac{224}{130}$. This is not too much. This kind of student is possible.

- If Alice has 1 small share and 3 large shares then she has $> \frac{53}{130} + 3 \times \frac{59}{130} = \frac{230}{130}$. This is not possible.

So Alice must have at least 1 large share and at least 2 small shares. The following scenario is possible:

- Four 4-students get 3 small shares and 1 large share.
- Three 4-students get 2 small shares and 2 large shares.

We leave it to the reader to show why the INT method cannot be used to obtain the upper bound. We need more information. $\frac{1}{2} = \frac{65}{130}$ is in the middle of the interval $(\frac{59}{130}, \frac{71}{130})$. The number of shares in $(\frac{59}{130}, \frac{65}{130})$ and $(\frac{65}{130}, \frac{71}{130})$ is the same by buddying.

The following picture captures what we know about the 4-shares:

$$(\ 18 \ 4\text{-shs} \ )[\ 0\ ](\ z \ 4\text{-shs} \ |\ z \ 4\text{-shs} \ )$$
$$\frac{53}{130} \qquad \frac{54}{130} \quad \frac{59}{130} \qquad \frac{65}{130} \qquad \frac{71}{130}$$

We define the following intervals:

- $I_1 = (\frac{53}{130}, \frac{54}{130})$ ($|I_1| = 18$);
- $I_2 = (\frac{59}{130}, \frac{65}{130})$;
- $I_3 = (\frac{65}{130}, \frac{71}{130})$ ($|I_2| = |I_3| = 5$).

This set of intervals does not quite capture all possible shares since $\frac{65}{130}$ is not in I_2 or I_3. We will address this point later in Convention 9.5. For now, we will put half of the $\frac{65}{130}$ shares in I_2 and half in I_3 (there will be an even number of them since $\frac{65}{130} = \frac{1}{2}$ and pieces of size $\frac{1}{2}$ have buddies of size $\frac{1}{2}$).

We need a finer classification of 4-students. We need to know how many shares from I_1, I_2, and I_3 a 4-student has.

Notation 9.4.

(1) If $1 \leq i \leq 3$ then *an I_i-share* is a share from I_i.
(2) Let $1 \leq j_1 \leq j_2 \leq j_3 \leq j_4 \leq 3$. A (j_1, j_2, j_3, j_4)-*student* is a student who has an I_{j_1}-share, an I_{j_2}-share, an I_{j_3}-share, and an I_{j_4}-share. The j's could be equal. We also refer to such a student as *a student of type* (j_1, j_2, j_3, j_4).

(3) y_{j_1,j_2,j_3,j_4} is the number of students of type (j_1, j_2, j_3, j_4).

(4) We often show that a student is impossible by showing that there is a $\beta \le \frac{m}{s}$ such that the student gets $< \beta$ muffins. If $\beta = \frac{m}{s}$ then we just barely showed that the student is impossible. This is usually a good sign. In these cases we put a * on it. Similarly, for showing that a student has too much muffins.

(5) Note that this notation can be extended to W-students and L intervals $I_1, \ldots, I_L$. Every W-student is a $(j_1, \ldots, j_W,)$-student for some $1 \le j_1 \le \cdots \le j_W$ (and again some of the j_i's may be equal). When dealing with 2-students, we will use z_{j_1,j_2} to denote the number of such students. (This won't come up until Section 11.3.)

Some types of students are impossible. For example, a $(1, 1, 1, 1)$-student has

$$< 4 \times \frac{54}{130} = \frac{216}{130} < \frac{230}{130}.$$

Similarly, a $(3, 3, 3, 3)$-student will have too much.

An interesting case is that of a $(1, 1, 2, 3)$-student. She will have

$$> 2 \times \frac{53}{130} + \frac{59}{130} + \frac{65}{130} = \frac{230}{130} *.$$

Although the student's I_3-share could be $\frac{65}{130}$, the students I_1-shares are both $> \frac{53}{130}$. Hence the student gets strictly more than $\frac{230}{130}$. Because the bound is tight we will put a * on the inequality. Recall that the * is a sign we are on the right track.

We now return to the problem of $\frac{65}{130}$ not being in any interval. There *can* be shares of that size. So how is our proof going to work? We describe a rigorous convention that we use here and throughout the rest of the book.

Convention 9.5. We state how having shares of size $\frac{65}{130}$ is not a problem for this proof; however, the points we make apply to many later proofs. We will later refer to this convention rather than restate the general points made obvious by this example.

- Above we showed that there are no (1, 1, 2, 3)-students since

$$2 \times \frac{53}{130} + \frac{59}{130} + \frac{65}{130} = \frac{230}{130}.$$

Since the I_1 shares are $> \frac{53}{130}$ it is not a problem that the I_3 share is $\geq \frac{65}{130}$. *Takeaway for the future*: the fact that there really could be shares of size $\frac{65}{130}$ does not matter if there are other shares that are endpoints of open intervals. This is a common case. There will be a case where this matters in Theorem 12.4.
- Above we claimed that $|I_2| = |I_3| = 5$. We obtain this by taking the shares of size $\frac{65}{130}$ and arbitrarily assigning half to I_2 and half to I_3. Recall that there is an even number of such shares since they buddy each other.

We determine which students are possible.

Claim: *The following are the only types of students who are possible*:

(1) (1, 1, 1, 3) ($y_{1,1,1,3}$ *denotes the number of such students*).
(2) (1, 1, 2, 2) ($y_{1,1,2,2}$ *denotes the number of such students*).

Proof of Claim:
We establish that some students are impossible.
  A (1, 1, 1, 2)-student has $< 3 \times \frac{54}{130} + \frac{65}{130} = \frac{227}{130} < \frac{230}{130}$. Hence there are no (1, 1, 1, 2)-students.
  A (1, 1, 2, 3)-student has $> 2 \times \frac{53}{130} + \frac{59}{130} + \frac{65}{130} = \frac{230}{130}*$. Hence there are no (1, 1, 2, 3)-students.
  A (1, 2, 2, 2)-student has $> \frac{53}{130} + 3 \times \frac{59}{130} = \frac{230}{130}*$. Hence there are no (1, 2, 2, 2)-students.
  The result follows from the set of impossible students.

End of Proof of Claim
 The (1, 1, 1, 3)-students do not use any I_2-shares. The (1, 1, 2, 2)-students each use two I_2-shares. Hence $|I_2| = 2y_{1,1,2,2}$.

The (1, 1, 1, 3)-students use one I_3-share. The (1, 1, 2, 2)-students do not use any I_3-shares. Hence $|I_3| = y_{1,1,1,3}$.

Since $|I_2| = |I_3|$:

$$2y_{1,1,2,2} = y_{1,1,1,3}. \tag{9.1}$$

Since $s_4 = 7$:

$$y_{1,1,1,3} + y_{1,1,2,2} = 7. \tag{9.2}$$

By substituting the expression for $y_{1,1,1,3}$ from Eq. (9.1) into Eq. (9.2) we obtain

$$3y_{1,1,2,2} = 7$$

$$y_{1,1,2,2} = \frac{7}{3}.$$

Recall that $y_{1,1,2,2}$ is the number of (1, 1, 2, 2)-students and hence is a natural. We can't have $\frac{7}{3}$ students of any type (unless the tables are turned and the muffins begin to cut up the student). Hence we have a contradiction. □

We show how one could *derive* the upper bound $f(23, 13) \le \frac{53}{130}$. Let α be the upper bound. We derive conditions on α that will make the proof of $f(23, 13) \le \alpha$ work. We assume $\alpha > \frac{1}{3}$. Using the V-conjecture, we guess everyone is either a 3-student or a 4-student.

The reasoning used in the proof above, that there are eighteen 3-shares and twenty-eight 4-shares, *did not use that the goal was* $\frac{53}{130}$. Hence we can use that reasoning. We have the following picture, though we do not know x or y.

$$(\text{ 28 4-shs })[\ 0\ ](\text{ 18 3-shs }\ )$$
$$\alpha \qquad\quad x \quad y \qquad\quad 1 - \alpha$$

What are x and y?

- x is the least number such that every 4-share is $< x$. Hence $3\alpha + x = \frac{230}{130}$, so $x = \frac{230}{130} - 3\alpha$.
- y is the largest number such that every 3-share is $> y$. Hence $2(1 - \alpha) + y = \frac{230}{130}$, so $y = 2\alpha - \frac{30}{130}$.

The following picture captures what we know.

$$\begin{array}{cccc} (\ 28\ \text{4-shs}\ )[& 0 &](& 18\ \text{3-shs}\ ) \\ \alpha & \frac{230}{130} - 3\alpha & 2\alpha - \frac{30}{130} & 1 - \alpha \end{array}$$

Since $[\frac{230}{130} - 3\alpha, 2\alpha - \frac{30}{130}] = \emptyset$, by buddying, $[\frac{160}{130} - 2\alpha, 3\alpha - \frac{100}{130}] = \emptyset$. We will assume that this new empty interval is within the 4-shares. This guess is based on the fact that there are more 4-shares than 3-shares. The following heuristic has always worked: when deciding which of the two intervals the buddy of the empty interval will be within, guess the interval with more shares.

The 4-shares:

$$\begin{array}{cccccc} (\ 18\ \text{4-shs}\ )[& 0 &](& 10\ \text{4-shs}\ )[& 0 &] \\ \alpha & \frac{160}{130} - 2\alpha & 3\alpha - \frac{100}{130} & \frac{230}{130} - 3\alpha & 2\alpha - \frac{30}{130} \end{array}$$

The 3-shares:

$$\begin{array}{cc} (\ & 18\ \text{3-shs}\ ) \\ 2\alpha - \frac{30}{130} & 1 - \alpha \end{array}$$

The interval $(3\alpha - \frac{100}{130}, \frac{230}{130} - 3\alpha)$ has $\frac{1}{2}$ is in the middle interval. By buddying, this interval will have the same number of shares to the left and to the right of $\frac{1}{2}$. The following picture captures what we know about the 4-shares (we do not know z).

$$\begin{array}{cccccc} (\ 18\ \text{4-shs}\ )[& 0 &](& z\ \text{4-shs}\ | & z\ \text{4-shs}\ ) \\ \alpha & \frac{160}{130} - 2\alpha & 3\alpha - \frac{100}{130} & \frac{65}{130} & \frac{230}{130} - 3\alpha \end{array}$$

We define the following intervals:

- $I_1 = (\alpha, \frac{160}{130} - 2\alpha)$ ($|I_1| = 18$);
- $I_2 = (3\alpha - \frac{100}{130}, \frac{65}{130})$;
- $I_3 = (\frac{65}{130}, \frac{230}{130} - 3\alpha)$ ($|I_2| = |I_3| = 5$).

There are 15 possible types of 4-students: $(1, 1, 1, 1)$, $(1, 1, 1, 2), \ldots, (3, 3, 3, 3)$. For each type we can find what values of α makes this type of student impossible. We give two examples.

(i) $(1, 1, 1, 1)$-student Alice is impossible if one of the following occurs:
- Alice has $< \frac{230}{130}$, so we need $4 \times (\frac{160}{130} - 2\alpha) < \frac{230}{130}$ which is equivalent to $\alpha > \frac{41}{104}$.
- Alice has $> \frac{230}{130}$, so we need $4\alpha > \frac{230}{130}$ which is equivalent to $\alpha > \frac{23}{52}$.

We view this calculation as making both $\frac{41}{104}$ and $\frac{23}{52}$ *candidates* for α.

(ii) $(1, 1, 2, 3)$-student Bob is impossible if one of the following occurs:
- Bob has $< \frac{230}{130}$, so we need $2 \times (\frac{160}{130} - 2\alpha) + \frac{65}{130} + \frac{230}{130} - 3\alpha < \frac{230}{130}$ which is equivalent to $\alpha > \frac{55}{130}$.
- Bob has $> \frac{230}{130}$, so we need $2\alpha + (3\alpha - \frac{100}{130}) + \frac{65}{130} > \frac{230}{130}$ which is equivalent to $\alpha > \frac{53}{130}$.

We view this calculation as making both $\frac{55}{130}$ and $\frac{53}{130}$ *candidates* for α. Note that these are merely candidates; some of them will not work.

To actually find the smallest α, find all 30 α's (two for each possible student types) and do a binary search to find the smallest one that works. In practice many of the candidates you generate are the same. Hence this method is not computationally intensive.

9.3. Exercises on Verifying

Exercise 9.6. Prove each of the following statements of the form $f(m, s) \leq \alpha$ using the techniques of this chapter. Alternatively, try to not look at the upper bounds and instead derive it.

(1) $f(23, 14) \leq \frac{17}{42}$.
(2) $f(43, 16) \leq \frac{50}{112}$.
(3) $f(33, 20) \leq \frac{49}{120}$.
(4) $f(37, 21) \leq \frac{103}{252}$.
(5) $f(59, 14) \leq \frac{131}{280}$.

Note 9.7. All of the upper bounds in Exercise 9.6 are actually lower bounds. So, for example, we know $f(59, 14) = \frac{131}{280}$.

9.4. The VMID Program

The proof that $f(23, 13) = \frac{53}{130}$ (Theorem 9.3) and the proofs in Exercise 9.6 all go as follows:

(1) Assume, by way of contradiction, that there is an (m, s)-procedure that shows $f(m, s) > \alpha$.
(2) Let $V = \lceil \frac{2m}{s} \rceil$. Assume the V-conjecture, so everyone is either a V-student or a $(V - 1)$-student. This assumption will be correct.
(3) Find out how many $(V - 1)$-students, V-students, $(V - 1)$-shares, and V-shares there are.
(4) Find the intervals that contain the $(V - 1)$-shares and the V-shares. If they are not disjoint then quit and output **DK**. (In this case it is likely that the FC method or the Half method can be used to prove the upper bound.)
(5) Use buddying to get that the W-shares (W will be $V - 1$ if there are more $(V - 1)$-shares, and V if there are more V-shares) are split into two intervals.
(6) (This step is where this method diverges from the INT method.) One of those two intervals has $\frac{1}{2}$ smack dab in the middle of it. We now create two intervals—the left half and the right half. These two intervals have the same number of shares in them.
(7) Determine which types of students are possible.
(8) Use that information to form a set of linear equations. If the system has no $\mathbb{N}$-solution then output **Yes** (so $f(m, s) \le \alpha$ has been verified) else output **DK**. (See Definition 5.5 to remind yourself that an $\mathbb{N}$-solution is a solution where all of the variables are in $\mathbb{N}$. Oh, looks like you don't have to remind yourself since I just did.)

We call this *the VMID method*.

9.5. Algorithms for MID: You Write It

The algorithms we ask you to write can use the algorithms in Chapter 6.

Exercise 9.8.

(1) Let VMID take as input m, s, α and try to verify $f(m, s) \leq \alpha$ via the MID method. Write an algorithm for VMID. It should output (1) **Yes** if $f(m, s) \leq \alpha$ can be proven with the VMID method, and (2) **DK** if $f(m, s) \leq \alpha$ cannot be established with the VMID method.

(2) Let MID take as input (m, s), derive good candidates α for the MID method, and then do a binary search to find the smallest α such that VMID$(m, s, \alpha) = $ **Yes**. If no candidate works output 1, indication that no upper bound can be derived using the MID method.

9.6. The MID Theorem

Definition 9.9. MID(m, s) is defined as follows:

(1) If s divides m then output 1.
(2) If not then run the program MID(m, s) and output its output.

Theorem 9.10. *If $m \leq s$ then $f(m, s) \leq $ MID(m, s).*

9.7. Program and Progress

Using the techniques presented so far we have the following attempt at an algorithm to find $f(m, s)$:

(1) Input(m, s).
(2) α is the min of

$$\{\text{FC}(m, s), \text{Half}(m, s), \text{INT}(m, s), \text{MID}(m, s)\}.$$

(3) Run FINDPROC(m, s, α). If it outputs a procedure P then output α, else output **DK**.

There are 3520 pairs (m, s) we are considering (see Chapter 3). There were 111 pairs that neither FC nor Half nor INT were able to solve, but MID was. Here are the full statistics so far. When we state that (say) for

329 cases $f(m, s) = \text{Half}(m, s)$, it is implicit that the prior techniques (in the case of Half its just FC) did not obtain the upper bound.

- For 2301 of them, $f(m, s) = \text{FC}(m, s)$. This is $\sim 65.37\%$.
- For 329 of them, $f(m, s) = \text{Half}(m, s)$. This is $\sim 9.35\%$.
- For 186 of them, $f(m, s) = \text{INT}(m, s)$. This is $\sim 5.28\%$.
- For 111 of them, $f(m, s) = \text{MID}(m, s)$. This is $\sim 3.15\%$.
- For 593 of them, none of FC, Half, INT, or MID suffices to find $f(m, s)$. This is $\sim 16.84\%$.

Chapter 10

The Easy Buddy–Match Method

10.1. Recap and Goals for This Chapter

From Chapters 4, 6, and 8 we know that, for $m \geq s$,

$$f(m, s) \leq \min\{\text{FC}(m, s), \text{Half}(m, s), \text{INT}(m, s), \text{MID}(m, s)\}.$$

Just for now we will refer to the min of those quantities as $\min f(m, s)$. Is it the case that, for all $m \geq s$, $f(m, s) = \min f(m, s)$? No. The counterexample with the smallest s is $f(16, 15)$; however it is more illustrative to show the following:

$$f(13, 12) \leq \frac{1}{3},$$

$$f(14, 13) \leq \frac{9}{26},$$

$$f(29, 27) \leq \frac{37}{108}.$$

(The first two results can be obtained by MID.)

The proof of the upper bounds uses a technique, which we call *easy buddy–match* (*EBM*). We develop an algorithm EBM(m, s) which, given m, s, outputs an α such that $f(m, s) \leq \alpha$. It only works when there are 2-shares (so $V = 3$). The key new idea is that if Alice is a 2-student and she has a share of size x, then her other share has size $\frac{m}{s} - x$. This is called *matching*. It is similar to buddying where, if there is a share of size x, there is a share of size $1 - x$.

Exercise 10.1. For $(m, s) = (16, 15), (17, 16), (19, 18), (29, 27)$:

(1) Compute $\alpha = \operatorname{minf}(m, s)$.
(2) Compute FINDPROC(m, s, α). (You should get a **DK** which means that $\operatorname{minf}(m, s)$ is unlikely to be $f(m, s)$.)

10.2. Matching

Recall the following scenario.

> *Assume that muffin M is cut into 2 pieces. Let x be a piece from M. M's other piece is of size $1 - x$. We call this other piece the buddy of x.*

Hence

> if (a, b) has c shares then
> $(1 - b, 1 - a)$ has c shares by buddying
> (also true for $[a, b]$ and $[1 - b, 1 - a]$).

We introduce a similar concept which we call *matching*.

> *Assume that Alice is a 2-student. Let x be one of Alice's shares. Alice's other share is of size $\frac{m}{s} - x$. We call this other piece the match of x.*

> If (a, b) has c 2-shares then
> $(\frac{m}{s} - b, \frac{m}{s} - a)$ has c 2-shares by matching
> (also true for $[a, b]$ and $[\frac{m}{s} - b, \frac{m}{s} - a]$).

Definition 10.2. Let x be a 2-share. *The match of x*, denoted $M(x)$, is $\frac{m}{s} - x$. This definition extends naturally to sets of 2-shares. We will be using it on intervals of 2-shares. We write $M(a, b)$ rather than the more proper $M((a, b))$. Similarly, we use $M[a, b]$ rather than $M([a, b])$. M is a bijection (see Appendix A for the definition of a bijection). Note that **we cannot apply M to a non-2-share!**

Since M is a bijection, if (a, b) is an interval contained in the 2-shares then (a, b) and $M(a, b)$ have the same number of shares. Why stop here? Recall that B (buddying) is a bijection. Hence (a, b), $M(a, b)$, and $B(M(a, b))$ are the same number of shares. Why stop here? Can we apply M? Be careful. In order to apply M we would need $B(M(a, b))$ to be contained in the 2-shares. Assume that $B(M(a, b))$ is contained in the 2-shares. Then I can apply M to obtain that (a, b), $M(a, b)$, $B(M(a, b))$, and $M(B(M(a, b))$ all have the same number of shares. We can now apply B to obtain that ... you get the idea. We want to continue this process as long as possible. When are we forced to stop? When we need to apply M but the interval we want to apply it to has non-2-shares. We formalize this with the definition of a buddy–match sequence, following the definition of 2-*share region*.

Definition 10.3. The 2-*share region* is a region that cannot have non-2-shares. Note that this includes an interval that we know has no shares.

Definition 10.4. A *buddy–match sequence* is a sequence of intervals $M_0, B_0, M_1, B_1, \ldots$ (the sequence is finite) such that (1) we apply B to M_i to get B_i, (2) M_i is contained in the 2-share region (so the next step makes sense), (3) we apply M to B_i to get M_{i+1}.

Recall the following notation.

Notation 10.5. If (a, b) is an interval then $|(a, b)|$ is the number of shares in that interval. Note that this is not standard.

10.3. Thirteen Muffins, Twelve Students

Theorem 10.6. $f(13, 12) = \frac{1}{3}$.

Proof. We show that $f(13, 12) \leq \frac{1}{3}$. We leave the proof that $f(13, 12) \geq \frac{1}{3}$ to the reader. Alternatively, the reader can run FINDPROC$(13, 12, \frac{1}{3})$ or use Theorem 4.5.

Assume, by way of contradiction, that there is a $(13, 12)$-procedure with smallest piece $> \frac{1}{3}$. By Theorem 2.6 every muffin is cut into exactly 2 pieces. Hence there are 26 pieces. We leave it to the reader to show that there are ten 2-students, two 3-students, twenty 2-shares, six 3-shares, and that the following picture captures what we know:

$$(\text{ 6 3-shs })(\text{ 20 2-shs })$$
$$\frac{4}{12} \qquad \frac{5}{12} \qquad \frac{8}{12}$$

We use a buddy–match sequence to find intervals that cover the 2-shares, which will lead to a contradiction.

- $(\frac{4}{12}, \frac{5}{12})$ contains 6 shares.
- $B(\frac{4}{12}, \frac{5}{12}) = (\frac{7}{12}, \frac{8}{12})$ contains 6 shares. Since $(\frac{7}{12}, \frac{8}{12})$ is contained in the 2-share region, we can apply M to it.
- $M(\frac{7}{12}, \frac{8}{12}) = (\frac{5}{12}, \frac{6}{12})$ contains 6 shares.
- $B(\frac{5}{12}, \frac{6}{12}) = (\frac{6}{12}, \frac{7}{12})$ contains 6 shares. Since $(\frac{6}{12}, \frac{7}{12})$ is contained in the 2-share region, we can apply M to it.
- $M(\frac{6}{12}, \frac{7}{12}) = (\frac{6}{12}, \frac{7}{12})$ contains 6 shares.
- $B(\frac{6}{12}, \frac{7}{12}) = (\frac{5}{12}, \frac{6}{12})$ contains 6 shares. Since $(\frac{5}{12}, \frac{6}{12})$ is contained in the 2-share region, we can apply M to it.
- $M(\frac{5}{12}, \frac{6}{12}) = (\frac{7}{12}, \frac{8}{12})$ contains 6 shares.
- $B(\frac{7}{12}, \frac{8}{12}) = (\frac{4}{12}, \frac{5}{12})$ contains 6 shares. Since $(\frac{4}{12}, \frac{5}{12})$ overlaps the 3-shares region we cannot apply M to it; we stop.
- There are no shares of size $\frac{4}{12}$. Since $B(\frac{4}{12}) = \frac{8}{12}$, there are no shares of size $\frac{8}{12}$. Similarly, there are no shares of the size of any the endpoints of the open intervals above. In particular, there are no shares of sizes $\frac{6}{12}, \frac{7}{12}$, or $\frac{8}{12}$.

The intervals $(\frac{5}{12}, \frac{6}{12})$, $(\frac{6}{12}, \frac{7}{12})$, and $(\frac{7}{12}, \frac{8}{12})$ each have 6 shares. Hence:

$$\left| \left(\frac{5}{12}, \frac{6}{12}\right) \cup \left(\frac{6}{12}, \frac{7}{12}\right) \cup \left(\frac{7}{12}, \frac{8}{12}\right) \right| = 18$$

and that there are no shares of sizes $\frac{5}{12}, \frac{6}{12}, \frac{7}{12}, \frac{8}{12}$. Hence there are 18 shares in $(\frac{5}{12}, \frac{8}{12})$ which is the interval that contains all the 2-shares. But there are 20 2-shares. This is a contradiction. □

The high-level view is that we used buddy–match to get a count of the 2-shares and it contradicted the real count.

10.4. Fourteen Students, Thirteen Muffins

Theorem 10.7. $f(14, 13) = \frac{9}{26}$.

Proof. We show that $f(14, 13) \le \frac{9}{26}$. We leave the proof that $f(14, 13) \ge \frac{9}{26}$ to the reader. Alternatively, the reader can run FINDPROC(14, 13, $\frac{9}{26}$).

Assume, by way of contradiction, that there is a (14, 13)-procedure with smallest piece $> \frac{9}{26}$. By Theorem 2.6, every muffin is cut into exactly 2 pieces. Hence there are 28 pieces. We leave it to the reader to show that there are eleven 2-students, two 3-students, twenty-two 2-shares, six 3-shares, and that the following picture captures what we know:

$$(\ 6 \text{ 3-shs })[\ 0 \ ](\qquad 22 \text{ 2-shs} \qquad)$$
$$\frac{9}{26} \qquad\quad \frac{10}{26} \ \frac{11}{26} \qquad\qquad\qquad \frac{17}{26}$$

We use a buddy–match sequence to find intervals that cover the entire interval which will lead to a contradiction.

- $(\frac{9}{26}, \frac{11}{26})$ has 6 shares. (The interval $(\frac{9}{26}, \frac{10}{26})$ has all 6 shares; however, it turns out to be easier to view $(\frac{9}{26}, \frac{11}{26})$ as having all six 3-shares, which it does.)
- $B(\frac{9}{26}, \frac{11}{26}) = (\frac{15}{26}, \frac{17}{26})$ contains 6 shares. Since $(\frac{15}{26}, \frac{17}{26})$ is contained in the 2-share region, we can apply M to it.
- $M(\frac{15}{26}, \frac{17}{26}) = (\frac{11}{26}, \frac{13}{26})$ contains 6 shares.
- $B(\frac{11}{26}, \frac{13}{26}) = (\frac{13}{26}, \frac{15}{26})$ contains 6 shares. Since $(\frac{13}{26}, \frac{15}{26})$ is contained in the 2-share region, we can apply M to it.
- $M(\frac{13}{26}, \frac{15}{26}) = (\frac{13}{26}, \frac{15}{26})$ contains 6 shares.
- $B(\frac{13}{26}, \frac{15}{26}) = (\frac{11}{26}, \frac{13}{26})$ contains 6 shares. Since $(\frac{11}{26}, \frac{13}{26})$ is contained in the 2-share region, we can apply M to it.
- $M(\frac{11}{26}, \frac{13}{26}) = (\frac{15}{26}, \frac{17}{26})$ contains 6 shares.
- $B(\frac{15}{26}, \frac{17}{26}) = (\frac{9}{26}, \frac{11}{26})$ contains 6 shares. Since $(\frac{9}{26}, \frac{11}{26})$ overlaps the 3-shares region, we cannot apply M to it; we stop.
- Since there are no shares of sizes $\frac{9}{26}$ or $\frac{11}{26}$, the above items also prove there are no shares of sizes $\frac{13}{26}, \frac{15}{26}$, or $\frac{17}{26}$.

Hence

$$\left| \left(\frac{9}{26}, \frac{11}{26}\right) \cup \left(\frac{11}{26}, \frac{13}{26}\right) \cup \left(\frac{13}{26}, \frac{15}{26}\right) \cup \left(\frac{15}{26}, \frac{17}{26}\right) \right| = 24$$

and that there are no shares of sizes $\frac{11}{26}, \frac{13}{26}, \frac{17}{26}$. Hence $(\frac{9}{26}, \frac{17}{26})$ has 24 shares. All of the shares are in this interval, so there are 24 shares. but there are 28 shares, which is a contradiction. □

Note the following difference:

- The proof that $f(13, 12) \le \frac{1}{3}$ looks at how many 2-shares there are.
- The proof that $f(14, 13) \le \frac{9}{26}$ looks at how many shares there are.

10.5. Twenty-Nine Muffins, Twenty-Seven Students

Theorem 10.8. $f(29, 27) = \frac{37}{108}$.

Proof. We show that $f(29, 27) \le \frac{37}{108}$. We leave the proof that $f(29, 27) \ge \frac{37}{108}$ to the reader. Alternatively, the reader can run FINDPROC$(29, 27, \frac{37}{108})$.

Assume, by way of contradiction, that there is a $(29, 27)$-procedure with smallest piece $> \frac{37}{108}$. By Theorem 2.6, every muffin is cut into exactly 2 pieces; hence there are 58 pieces. We leave it to the reader to show that there are twenty-three 2-students, four 3-students, forty-six 2-shares, twelve 3-shares, and that the following picture captures what we know:

$$(\ 12 \text{ 3-shs })[\ 0 \ ](\ 46 \text{ 2-shs })$$
$$\frac{37}{108} \qquad\qquad \frac{42}{108} \ \frac{45}{108} \qquad\qquad \frac{71}{108}$$

We use a buddy–match sequence to find a useful empty interval.

(1) $[\frac{42}{108}, \frac{45}{108}]$ is empty.
(2) $B[\frac{42}{108}, \frac{45}{108}] = [\frac{63}{108}, \frac{66}{108}]$ is empty. Since $[\frac{63}{108}, \frac{66}{108}]$ is contained in the 2-share region, so we can apply M to it.
(3) $M[\frac{63}{108}, \frac{66}{108}] = [\frac{50}{108}, \frac{53}{108}]$ is empty.
(4) $B[\frac{50}{108}, \frac{53}{108}] = [\frac{55}{108}, \frac{58}{108}]$ is empty. Since $[\frac{55}{108}, \frac{58}{108}]$ is contained in the 2-share region, so we can apply M to it.
(5) $M[\frac{55}{108}, \frac{58}{108}] = [\frac{58}{108}, \frac{61}{108}]$ is empty.
(6) $B[\frac{58}{108}, \frac{61}{108}] = [\frac{47}{108}, \frac{50}{108}]$ is empty. Since $[\frac{47}{108}, \frac{50}{108}]$ is contained in the 2-share region, so we can apply M to it.
(7) $M[\frac{47}{108}, \frac{50}{108}] = [\frac{66}{108}, \frac{69}{108}]$ is empty.

(8) $B[\frac{66}{108}, \frac{69}{108}] = [\frac{39}{108}, \frac{42}{108}]$ is empty. Since $[\frac{39}{108}, \frac{42}{108}]$ overlaps the 3-shares region, we cannot apply M to it; we stop.

(9) None of the endpoints have shares.

Since $[\frac{39}{108}, \frac{42}{108}]$ is empty and $[\frac{42}{108}, \frac{45}{108}]$ is empty, then $[\frac{39}{108}, \frac{45}{108}]$ is empty. We could deduce many other closed intervals of length $\frac{6}{108}$ are empty; however, all we need is $[\frac{39}{108}, \frac{42}{108}]$. We will use this interval being empty in a crucial part of the next buddy–match sequence.

The following picture captures what we know:

$$(\text{ 12 3-shs })[\;0\;](\text{ 46 2-shs })$$
$$\underset{\frac{37}{108}}{} \qquad\qquad \underset{\frac{39}{108}}{}\;\underset{\frac{45}{108}}{} \qquad\qquad \underset{\frac{71}{108}}{}$$

It is important that the gap has been expanded from $[\frac{42}{108}, \frac{45}{108}]$ to $[\frac{39}{108}, \frac{45}{108}]$. Why? In the following buddy–match sequence, at step 8, we will be able to do a match step, where we would not have been able to if we had the smaller gap.

We use a buddy–match sequence to find intervals that cover the entire interval, which will lead to a contradiction.

(1) $(\frac{37}{108}, \frac{45}{108})$ has 12 shares.

(2) $B(\frac{37}{108}, \frac{45}{108}) = (\frac{63}{108}, \frac{71}{108})$ has 12 shares. Since $(\frac{63}{108}, \frac{71}{108})$ is contained in the 2-share region, we can apply M to it.

(3) $M(\frac{63}{108}, \frac{71}{108}) = (\frac{45}{108}, \frac{53}{108})$ has 12 shares.

(4) $B(\frac{45}{108}, \frac{53}{108}) = (\frac{55}{108}, \frac{63}{108})$ has 12 shares. Since $(\frac{55}{108}, \frac{63}{108})$ is contained in the 2-share region, we can apply M to it.

(5) $M(\frac{55}{108}, \frac{63}{108}) = (\frac{53}{108}, \frac{61}{108})$ has 12 shares.

(6) $B(\frac{53}{108}, \frac{61}{108}) = (\frac{47}{108}, \frac{55}{108})$ has 12 shares. Since $(\frac{47}{108}, \frac{55}{108})$ is contained in the 2-share region, we can apply M to it.

(7) $M(\frac{47}{108}, \frac{55}{108}) = (\frac{61}{108}, \frac{69}{108})$ has 12 shares.

(8) $B(\frac{61}{108}, \frac{69}{108}) = (\frac{39}{108}, \frac{47}{108})$ has 12 shares. Since $(\frac{39}{108}, \frac{47}{108})$ is contained in the 2-share region, we can apply M to it. Note that if we still had the smaller gap of $[\frac{42}{108}, \frac{45}{108}]$ then we could not state that $(\frac{39}{108}, \frac{47}{108})$ was in the 2-share region.

(9) $M(\frac{39}{108}, \frac{47}{108}) = (\frac{69}{108}, \frac{77}{108})$ has 12 shares. (This looks odd since there are no shares in $[\frac{71}{108}, \frac{77}{108})$ but it is still true.)

(10) $B(\frac{69}{108}, \frac{77}{108}) = (\frac{31}{108}, \frac{39}{108})$ has 12 shares. Since $(\frac{31}{108}, \frac{39}{108})$ overlaps the 3-shares region, we cannot apply M to it; we stop.

(11) None of the endpoints have shares.

Each of the intervals in the next line has 12 shares. Hence the total number of shares is $5 \times 12 = 60$.

$$\left(\frac{37}{108}, \frac{45}{108}\right) \cup \left(\frac{45}{108}, \frac{53}{108}\right) \cup \left(\frac{53}{108}, \frac{61}{108}\right) \cup \left(\frac{61}{108}, \frac{69}{108}\right) \cup \left(\frac{69}{108}, \frac{77}{108}\right)$$

However there are 58 shares. This is a contradiction. $\square$

Exercise 10.9. Imitate the proofs of Theorems 10.6, 10.7, or and 10.8 to prove the following:

(1) $f(33, 31) \le \frac{21}{62}$.

(2) $f(35, 33) \le \frac{15}{44}$.

(3) $f(37, 34) \le \frac{1}{3}$.

(4) $f(39, 37) \le \frac{25}{74}$.

(5) $f(41, 39) \le \frac{53}{156}$.

(6) $f(44, 41) \le \frac{14}{41}$.

(7) $f(45, 43) \le \frac{29}{86}$.

(8) $f(46, 43) \le \frac{1}{3}$.

(9) $f(47, 45) \le \frac{61}{180}$.

(10) $f(49, 45) \le \frac{1}{3}$.

(11) $f(49, 46) \le \frac{31}{92}$.

(12) $f(50, 47) \le \frac{16}{47}$.

(13) $f(51, 49) \le \frac{33}{98}$.

(14) $f(53, 50) \le \frac{17}{50}$.

10.6. General Theorem

We prove a theorem about $f(3dk + a + d, 3dk + a)$. We will only look at $a \in \{1, \ldots, 3d\}$ because if $a \geq 3d + 1$ then we can use a larger k. We note now that we will not need a VEBM (Verify EBM) function; we go straight to EBM which will be a set of nice formulas.

Why $(3dk + a + d, 3dk + a)$? Originally, we were looking at $f(s + d, s)$ for fixed d. We were hoping for formulas like we had for $f(m, s)$ for fixed s. (Such formulas can be obtained from our work; however, it is not in the book. It is on the MUFFIN website.) The formulas depended on $m \pmod{3d}$. Hence we ended up looking at $(3dk + a + d, 3dk + a)$. More formally, given m, s we find d, k, a as follows:

(1) $d = m - s$.
(2) k is the largest number ≥ 0 such that $3dk < s$.
(3) $a = s - 3dk$.

We do not use these equations until we define the EBM function in Section 10.7.

We will soon prove a general theorem, Theorem 10.11, that gives formulas for upper bounds on $f(m, s)$. However, there is one case that is easier to do separately. That case is the following exercise.

Exercise 10.10. In this exercise, due to Yunseo Choi and Kevin Cong, we look at the case of $a = 2d$.

(1) Use the Floor–Ceiling theorem to get an upper bound on $f(3dk + 3d, 3dk + 2d)$.
(2) Show that the upper bound is also a lower bound.

Because of Exercise 10.10 we can ignore the $a = 2d$ case in the following theorem.

Theorem 10.11. *Let* $d, k \geq 1$, $a \in \{1, \ldots, 3d\} - \{2d\}$.

(1) *If* $2d + 1 \leq a \leq 3d$, *then*

$$f(3dk + a + d, 3dk + a) \leq \frac{dk + X}{3dk + a} = \frac{1}{3},$$

where $X = \frac{a}{3}$. (*This case is a generalization of Theorem* 10.6. *Plug* $d = 1, a = 3, k = 3$ *into this formula to obtain Theorem* 10.6.)

(2)
$$f(3dk + a + d, 3dk + a) \le \frac{dk + X}{3dk + a},$$

where $X = \frac{a}{2}$. *Only useful when* $a \in \{1, \ldots, d\}$ *since otherwise part* (1) *or part* (3) *gives a stronger result.* (*This case is a generalization of Theorem* 10.7. *Plug* $d = 1, a = 1, k = 4$ *into this formula to obtain Theorem* 10.7.)

(3)
$$f(3dk + a + d, 3dk + a) \le \frac{dk + X}{3dk + a},$$

where $X = \frac{a+d}{4}$. *Only useful when* $a \in \{d, \ldots, 2d - 1\}$ *since otherwise part* (1) *or part* (2) *gives a stronger result.* (*This case is a generalization of Theorem* 10.8. *Plug* $d = 2, a = 3, k = 4$ *into this formula to obtain Theorem* 10.8.)

Proof. We do the proof with parameter X and then break into three cases (Cases 3e.1–3e.3) corresponding to the three statements in the theorem. We assume $X \ge \frac{a}{3}$ so $\frac{dk+X}{3dk+a} \ge \frac{1}{3}$.

Assume, by way of contradiction, that there is a $(3dk + a + d, 3dk + a)$-procedure with smallest piece $> \frac{dk+X}{3dk+a} \ge \frac{1}{3}$. By Theorem 2.6, every muffin is divided into two pieces. Hence, there are $6dk + 2a + 2d$ pieces.

Case 1: If Alice gets ≥ 4 shares, then some share is

$$\le \frac{3dk + a + d}{3dk + a} \times \frac{1}{4} = \frac{0.75dk + 0.25a + 0.25d}{3dk + a} \le \frac{dk + X}{3dk + a}.$$

(The inequality follows from $X \ge \frac{a}{3} > \frac{a}{4}$ and $k \ge 1$.)

Case 2: If Alice gets 1 share then, since each share is $\le \frac{1}{2}$, she has

$$\le \frac{1}{2} < \frac{3dk + a + d}{3dk + a},$$

which is impossible.

Case 3: Everyone is a 2-student or a 3-student.

Let s_2 (s_3) be the number of 2-students (3-students). Since there are $6dk + 2a + 2d$ pieces, we have

$$2s_2 + 3s_3 = 6dk + 2a + 2d,$$

$$s_2 + s_3 = 3dk + a.$$

Hence $s_2 = 3dk + a - 2d$ and $s_3 = 2d$. So there are $6dk + 2a - 4d$ 2-shares and $6d$ 3-shares.

Case 3a: There is a 3-share $\geq \frac{dk+a+d-2X}{3dk+a}$. The remaining two 3-shares add up to

$$\leq \frac{3dk + a + d}{3dk + a} - \frac{dk + a + d - 2X}{3dk + a} = \frac{2dk + 2X}{3dk + a};$$

hence there is a 3-share

$$\leq \frac{2dk + 2X}{3dk + a} \times \frac{1}{2} = \frac{dk + X}{3dk + a}.$$

Case 3b: There is a 2-share $\geq \frac{2dk+a-X}{3dk+a}$. Its buddy is

$$\leq 1 - \frac{2dk + a - X}{3dk + a} = \frac{dk + X}{3dk + a}.$$

Case 3c: There is a 2-share $\leq \frac{dk+d+X}{3dk+a}$. Its match is

$$\geq \frac{3dk + a + d}{3dk + a} - \frac{dk + d + X}{3dk + a} = \frac{2dk + a - X}{3dk + a}.$$

Its buddy is

$$\leq 1 - \frac{2dk + a - X}{3dk + a} = \frac{dk + X}{3dk + a}.$$

Case 3d: There is a 3-share $\leq \frac{dk+X}{3dk+a}$. This case is obvious.

Case 3e: All the 3-shares are in $(\frac{dk+X}{3dk+a}, \frac{dk+a+d-2X}{3dk+a})$ and all the 2-shares are in $(\frac{dk+d+X}{3dk+a}, \frac{2dk+a-X}{3dk+a})$.

The following picture captures what we know:

$$(\quad 3s_3 \text{ 3-shs} \quad)[\quad\quad 0 \quad](\quad 2s_2 \text{ 2-shs} \quad)$$
$$\frac{dk+X}{3dk+a} \quad\quad \frac{dk+a+d-2X}{3dk+a} \quad \frac{dk+d+X}{3dk+a} \quad\quad \frac{2dk+a-X}{3dk+a}$$

(The reader can check that the assumption $X \geq \frac{a}{3}$ ensures that the interval of 2-shares and the interval of 3-shares do not intersect.)

We define the following:

$$M_0 = \left(\frac{dk+X}{3dk+a}, \frac{dk+d+X}{3dk+a}\right),$$

$$B_0 = B(M_0) = \left(\frac{2dk+a-d-X}{3dk+a}, \frac{2dk+a-X}{3dk+a}\right),$$

$$(\forall 0 \le i \le k-1)\left[M_i = M(B_{i-1})\right.$$
$$= \left.\left(\frac{dk+id+X}{3dk+a}, \frac{dk+(i+1)d+X}{3dk+a}\right)\right],$$

$$(\forall 0 \le i \le k-1)\left[B_i = B(M_i)\right.$$
$$= \left.\left(\frac{2dk+a-(i+1)d-X}{3dk+a}, \frac{2dk+a-id-X}{3dk+a}\right)\right].$$

The last $B(M)$ to be defined is $B_{k-1}(M_{k-1})$. We stop there since we *might* have an interval that overlaps the 3-share region, and hence cannot apply M to it.

We want that if $k \ge 2$, then B_{k-2}, and hence $B_0, \ldots, B_{k-3}$, are contained in the 2-share region; hence M can be applied to them. We want this since the definition of M_{k-1} needs that B_{k-2} is contained in the 2-share region. We need the left endpoint of B_{k-2} to be greater than or equal to the right endpoint of the 3-shares which we take to be $\frac{dk+a+d-2X}{3dk+a}$. Hence we need

$$\frac{dk+a+d-2X}{3dk+a} \le \frac{2dk+a-(k-1)d-X}{3dk+a},$$

which is equivalent to $1 \le 2$, which is true of course.

Since $|B_0| = 6d$ and the B and M functions are bijections, (1) all B_i's and M_i's have $6d$ shares, and (2) if x is an endpoint of any B_i or M_i then there are no shares of size x.

There are three subcases of Case 3 that correspond to the three statements in our theorem.

Case 3e.1 (Statement 1 of the theorem): We assume, by way of contradiction, that $2d+1 \le a \le 3d$ and $X > \frac{a}{3}$.

We show that
$$B_0 \cup \cdots \cup B_{k-1} \supseteq \left(\frac{dk+d+X}{3dk+a}, \frac{2dk+a-X}{3dk+a}\right).$$

(The right-hand side is the set of 2-shares.)
We want the left endpoint of B_{k-1} to be $\leq$ the right endpoint of the 2-shares which we take to be $\frac{dk+d+X}{3dk+a}$. Hence we need

$$\frac{2dk+a-kd-X}{3dk+a} \leq \frac{dk+a+d-2X}{3dk+a},$$

which is $X \leq d$. This is true since $X = \frac{a}{3}$ and $a \leq 3d$.
So we have

$$B_0 \cup \cdots \cup B_{k-1} \supseteq \left(\frac{dk+d+X}{3dk+a}, \frac{2dk+a-X}{3dk+a}\right).$$

(The right-hand side is the set of 2-shares.)

$$|B_0 \cup \cdots \cup B_{k-1}| \geq \left|\left(\frac{dk+d+X}{3dk+a}, \frac{2dk+a-X}{3dk+a}\right)\right|$$

$$6dk \geq 6dk + 2a - 4d$$

$$2d \geq a,$$

which is a contradiction.

Case 3e.2 (Statement 2 of the theorem): Assume, by way of contradiction, that $X > \frac{a}{2}$.

We show that
$$B_0 \cup \cdots \cup B_{k-1} = \left(\frac{dk+X}{3dk+a}, \frac{2dk+a-X}{3dk+a}\right).$$

(The right-hand side is the entire interval.)
We want the left endpoint of B_{k-1} to be $\leq$ the right endpoint of the 3-shares which is $\frac{dk+X}{3dk+a}$. Hence we need

$$\frac{2dk+a-kd-X}{3dk+a} \leq \frac{dk+X}{3dk+a},$$

which is $X \geq \frac{a}{2}$.

Since the last line is true, we have

$$B_0 \cup \cdots \cup B_{k-1} = \left(\frac{dk+X}{3dk+a}, \frac{2dk+a-X}{3dk+a}\right)$$

$$|B_0 \cup \cdots \cup B_{k-1}| = \left|\left(\frac{dk+X}{3dk+a}, \frac{2dk+a-X}{3dk+a}\right)\right|$$

$$6dk = 6dk + 2d + 2a$$

$$a + d = 0,$$

which is a contradiction since $d \geq 1$.

Case 3e.3 (Statement 3 of the theorem): Assume, by way of contradiction, that $X > \frac{a+d}{4}$. (Recall that $a \neq 2d$ which we use later.)

If $X > \frac{a}{2}$ then, by Case 3e.2, we already get a contradiction. Hence we assume $X \leq \frac{a}{2}$ during this case.

We show the picture that captures what we know:

$$\underset{\frac{dk+X}{3dk+a}}{(}\quad 3s_3\ 3\text{-shs}\quad \underset{\frac{dk+a+d-2X}{3dk+a}}{)[}\quad 0 \quad \underset{\frac{dk+d+X}{3dk+a}}{](}\quad 2s_2\ 2\text{-shs}\quad \underset{\frac{2dk+a-X}{3dk+a}}{)}$$

We use a buddy–match sequence to find a useful empty interval contained in the 3-share region.

$$M_0 = \left[\frac{dk+a+d-2X}{3dk+a}, \frac{dk+d+X}{3dk+a}\right],$$

$$B_0 = B(M_0) = \left[\frac{2dk+a-d-X}{3dk+a}, \frac{2dk-d+2X}{3dk+a}\right],$$

$$M_1 = M(B_0) = \left[\frac{dk+a+2d-2X}{3dk+a}, \frac{dk+2d+X}{3dk+a}\right],$$

$$B_1 = B(M_0) = \left[\frac{2dk+a-2d-X}{3dk+a}, \frac{2dk-2d+2X}{3dk+a}\right],$$

$$(\forall 0 \leq i \leq k-1)$$

$$\times \left[M_i = M(B_{i-1}) = \left[\frac{dk+a+(i+1)d-2X}{3dk+a}, \frac{dk+(i+1)d+X}{3dk+a}\right],\right.$$

$(\forall 0 \leq i \leq k-1)$
$$\times \left[B_i = B(M_i) = \left[\frac{2dk + a - (i+1)d - X}{3dk + a}, \frac{2dk - (i+1)d + 2X}{3dk + a} \right] \right..$$

We want B_{k-2}, and hence $B_0, \ldots, B_{k-3}$, to be contained in the 2-shares, so that M can be applied to them. Note that

$$B_{k-2} = \left[\frac{dk + a + d - X}{3dk + a}, \frac{dk + d - X}{3dk + a} \right].$$

We need

$$\frac{dk + a + d - 2X}{3dk + a} \leq \frac{dk + a + d - X}{3dk + a},$$

which is clearly true.

Since M_0 is empty, so are all of the M_i's and B_i's. So

$$B_{k-1} = \left[\frac{dk + a - X}{3dk + a}, \frac{dk + 2X}{3dk + a} \right] = \emptyset.$$

We want the B_{k-1} gap to merge with the $(\frac{dk+a+d-2X}{3dk+a}, \frac{dk+d-2X}{3dk+a})$ gap to form a larger gap. Hence we need

$$\frac{dk + X}{3dk + a} \leq \frac{dk + a - X}{3dk + a} \leq \frac{dk + a + d - 2X}{3dk + a} \leq \frac{dk + 2X}{3dk + a}.$$

The first inequality is equivalent to $X \leq \frac{a+d}{2}$. We have this since $X \leq \frac{a}{2}$.
The second inequality is equivalent to $X \leq d$. We have this since $X \leq \frac{a}{2}$ and $a \leq 2d$.
The third inequality is equivalent to $X \geq \frac{a+d}{4}$. We have this by hypothesis.

The following picture captures what we know:

$$\underset{\frac{dk+X}{3dk+a}}{(} \quad 3s_3 \text{ 3-shs} \quad \underset{\frac{dk+a-X}{3dk+a}}{)[} \quad 0 \quad \underset{\frac{dk+d+X}{3dk+a}}{](} \quad 2s_2 \text{ 2-shs} \quad \underset{\frac{2dk+a-X}{3dk+a}}{)}$$

We use a buddy–match sequence to find intervals that cover the entire interval, which will cause a contradiction. This is similar to the sequence in

Case 3e.2 except that, because of the bigger gap, we can go one step further.

$$M_0 = \left(\frac{dk+X}{3dk+a}, \frac{dk+X+d}{3dk+a}\right),$$

$$B_0 = B(M_0) = \left(\frac{2dk+a-d-X}{3dk+a}, \frac{2dk+a-X}{3dk+a}\right),$$

$(\forall 0 \leq i \leq k)$

$$\times \left[M_i = M(B_{i-1}) = \left(\frac{dk+id+X}{3dk+a}, \frac{dk+(i+1)d+X}{3dk+a}\right)\right],$$

$(\forall 0 \leq i \leq k)$

$$\times \left[B_i = B(M_i) = \left(\frac{2dk+a-(i+1)d-X}{3dk+a}, \frac{2dk+a-id-X}{3dk+a}\right)\right].$$

We want B_{k-1}, and hence $B_0, \ldots, B_{k-2}$ are contained in the 2-share region, so that M can be applied to them. Note that

$$B_{k-1} = \left(\frac{dk+a-X}{3dk+a}, \frac{dk+a \mid d \quad X}{3dk+a}\right).$$

We need

$$\frac{dk+a-X}{3dk+a} \leq \frac{dk+a-X}{3dk+a},$$

which is clearly true.

We want $M_0 \cup \cdots \cup M_k$ to cover the entire interval. Hence we need

$$\frac{2dk+a-X}{3dk+a} \leq \frac{dk+(k+1)d+X}{3dk+a},$$

which is $X \geq \frac{a-d}{2}$. We have this since $X \geq \frac{a+d}{4}$ and $a \leq 3d$.

Since $|B_0| = 6d$ and B, M are bijections, (1) all B_i's and M_i's have $6d$ shares, and (2) there are no shares that have the same size as the endpoints of any B_i or M_i. Hence

$$6dk + 2a + 2d = |M_0 \cup \cdots \cup M_k| = 6d(k+1),$$

which implies $a = 2d$. This is a contradiction. □

10.7. The Function EBM

We define a function EBM so that we can add it to the list of upper bounds on $f(m, s)$.

Definition 10.12. EBM(m, s) is defined as follows.

(1) If s divides m then EBM$(m, s) = 1$.
(2) If $\lceil \frac{2m}{s} \rceil \geq 4$ then EBM$(m, s) = 1$.
(3) Let:
   (a) $d = m - s$.
   (b) k be the largest number ≥ 0 such that $3dk < s$.
   (c) $a = s - 3dk$.
   (d) Note that $s = 3dk + a$ and $m = 3dk + a + d$.
(4) (a) If $a \in \{2d+1, \ldots, 3d\}$ then EBM$(m, s) = \frac{dK+X}{3dk+a}$ where $X = \frac{a}{3}$, so EBM$(m, s) = \frac{1}{3}$.
   (b) If $a = 2d$ then EBM$(m, s) = $ FC(m, s) by Exercise 10.10.
   (c) If $a \in \{1, \ldots, d\}$ then EBM$(m, s) = \frac{dK+X}{3dk+a}$ where $X = \min\{\frac{a}{2}\}$.
   (d) If $a \in \{d, \ldots, 2d-1\}$ then EBM$(m, s) = \frac{dK+X}{3dk+a}$ where $X = \min\{\frac{a+d}{4}\}$.

From Theorem 10.11, we have the following theorem.

Theorem 10.13. *For all $m \geq s$, $f(m, s) \leq $ EBM(m, s).*

10.8. Program and Progress

Using the techniques presented so far we have the following attempt at an algorithm to find $f(m, s)$:

(1) Input(m, s).
(2) α is the min of

$$\{\text{FC}(m, s), \text{Half}(m, s), \text{MID}(m, s), \text{INT}(m, s), \text{EBM}(m, s)\}.$$

(3) Run FINDPROC(m, s, α). If it outputs a procedure P then output α, else output **DK**.

There are 3520 pairs (m, s) we are considering (see Chapter 3). There were 198 pairs that neither FC nor Half nor INT nor MID able to solve, but EBM was. Here are the full statistics so far. When we state that (say) for 329 cases $f(m, s) = \text{Half}(m, s)$, it is implicit that the prior techniques (in the case of Half its just FC) did not obtain the upper bound.

- For 2301 of them, $f(m, s) = \text{FC}(m, s)$. This is $\sim 65.37\%$.
- For 329 of them, $f(m, s) = \text{Half}(m, s)$. This is $\sim 9.35\%$.
- For 186 of them, $f(m, s) = \text{INT}(m, s)$. This is $\sim 5.28\%$.
- For 111 of them, $f(m, s) = \text{MID}(m, s)$. This is $\sim 3.15\%$.
- For 240 of them, $f(m, s) = \text{EBM}(m, s)$. This is $\sim 6.82\%$.
- For 353 of them, none of FC, Half, INT, MID, or EBM suffices to find $f(m, s)$. This is $\sim 10.00\%$.

Chapter 11

The Hard Buddy–Match Method

11.1. Recap and Goals for This Chapter

From Chapters 4, 6, and 8–10 we know that, for $m \geq s$,

$$f(m, s) \leq \min\{FC(m, s), \text{Half}(m, s), \text{MID}(m, s), \text{INT}(m, s),$$
$$\text{EBM}(m, s)\}.$$

Just for now we will refer to the min of those quantities as $\text{minf}(m, s)$. Is it the case that, for all $m \geq s$, $f(m, s) = \text{minf}(m, s)$? No. The counterexample with the smallest s is $f(25, 22)$; however it is more illustrative to show the following:

$$f(59, 52) \leq \frac{9}{26} < \text{minf}(59, 52).$$

The proof of the upper bound uses a technique, which we call *hard buddy–match* (*HBM*). We develop an algorithm $HBM(m, s)$ which, given m, s, outputs an α such that $f(m, s) \leq \alpha$. It only works when $V = 3$.

Exercise 11.1. For $(m, s) = (25, 22), (33, 29), (34, 31), (59, 52)$:

(1) Compute $\alpha = \text{minf}(m, s)$.
(2) Compute $\text{FINDPROC}(m, s, \alpha)$. (You should get a **DK** which means that $\text{minf}(m, s)$ is unlikely to be $f(m, s)$.)

11.2. $f(59, 52) \leq \frac{9}{26}$

Theorem 11.2. $f(59, 52) = \frac{9}{26}$.

Proof. We leave the proof that $f(59, 52) \geq \frac{9}{26}$ to the reader. Alternatively, the reader can run FINDPROC(59, 52, $\frac{9}{26}$).

We use denominator 52 throughout. We restate our theorem as $f(59, 52) = \frac{18}{52}$.

Assume, by way of contradiction, that there is a $(59, 52)$-procedure with smallest piece $> \frac{18}{52}$. By Theorem 2.6, every muffin is cut into exactly 2 pieces. Hence there are 118 pieces. We leave it to the reader to show that there are thirty-eight 2-students, fourteen 3-students, seventy-six 2-shares, forty-two 3-shares, and that the following picture captures what we know:

$$(\ 42 \ 3\text{-shs} \ )[\ 0 \ ](\ 76 \ 2\text{-shs} \ )$$
$$\frac{18}{52} \qquad\qquad \frac{23}{52} \ \frac{25}{52} \qquad\qquad \frac{34}{52}$$

We use a buddy–match sequence to find more empty intervals.

(1) $\left[\frac{23}{52}, \frac{25}{52}\right]$ is empty.
(2) $B\left[\frac{23}{52}, \frac{25}{52}\right] = \left[\frac{27}{52}, \frac{29}{52}\right]$ is empty. Since $\left[\frac{27}{52}, \frac{29}{52}\right]$ is contained in the 2-share region, we can apply M to it.
(3) $M\left[\frac{27}{52}, \frac{29}{52}\right] = \left[\frac{30}{52}, \frac{32}{52}\right]$ is empty.
(4) $B\left[\frac{30}{52}, \frac{32}{52}\right] = \left[\frac{20}{52}, \frac{22}{52}\right]$ is empty. Since $\left[\frac{20}{52}, \frac{22}{52}\right]$ overlaps the 3-shares region, we cannot apply M; we stop.

The following picture captures what we know (we do not know z):

$$(\ z \ 3\text{-shs} \ )[\ 0 \ ](\ 42-z \ 3\text{-shs} \ )[\ 0 \ ]$$
$$\frac{18}{52} \qquad\qquad \frac{20}{52} \ \frac{22}{52} \qquad\qquad \frac{23}{52} \ \frac{25}{52}$$

$$(\ 34 \ 2\text{-shs} \ )[\ 0 \ ](\ 42-z \ 2\text{-shs} \ )[\ 0 \ ](\ z \ 2\text{-shs} \ )$$
$$\frac{25}{52} \qquad\qquad \frac{27}{52} \ \frac{29}{52} \qquad\qquad \frac{30}{52} \ \frac{32}{52} \qquad\qquad \frac{34}{52}$$

We want to know what z is. Hence we do a buddy–match sequence beginning with $\left(\frac{18}{52}, \frac{20}{52}\right)$, which has z shares, hoping it gets mapped to an interval with a known number of shares.

(1) $\left(\frac{18}{52}, \frac{20}{52}\right)$ has z shares.
(2) $B\left(\frac{18}{52}, \frac{20}{52}\right) = \left(\frac{32}{52}, \frac{34}{52}\right)$ has z shares (which we already knew).
(3) $M\left(\frac{32}{52}, \frac{34}{52}\right) = \left(\frac{25}{52}, \frac{27}{52}\right)$ has z shares. But we also know it has 34 shares. So $z = 34$.

The following picture captures what we know:

$$\begin{array}{cccccc} (\ 34\ 3\text{-shs}\ )[\ 0\ ](& 8\ 3\text{-shs}\ )[\ 0\ ] & & & & \\ \frac{18}{52} & \frac{20}{52} & \frac{22}{52} & & \frac{23}{52} & \frac{25}{52} \end{array}$$

$$\begin{array}{cccccc} (\ 34\ 2\text{-shs}\ )[\ 0\ ](& 8\ 2\text{-shs}\ )[\ 0\ ](& 34\ 2\text{-shs}\ ) & & & \\ \frac{25}{52} & \frac{27}{52} & \frac{29}{52} & \frac{30}{52} & \frac{32}{52} & \frac{34}{52} \end{array}$$

We use a buddy–match sequence to find symmetry.

(1) $B\left(\frac{18}{52}, \frac{20}{52}\right) = \left(\frac{32}{52}, \frac{34}{52}\right)$. It has no 3-shares so we can apply M.
(2) $M\left(\frac{32}{52}, \frac{34}{52}\right) = \left(\frac{25}{52}, \frac{27}{52}\right)$. This is symmetric around $\frac{26}{52} = \frac{1}{2}$ so $\left(\frac{18}{52}, \frac{20}{52}\right)$ is symmetric around $\frac{19}{52}$.

The following picture captures what we know about the 3-shares:

$$\begin{array}{ccccc} (\ 17\ 3\text{-shs}\ |\ 17\ 3\text{-shs}\ )[\ 0\ ](& 8\ 3\text{-shs}\ ) & & & \\ \frac{18}{52} & \frac{19}{52} & \frac{20}{52} & \frac{22}{52} & \frac{23}{52} \end{array}$$

We define the following intervals and use Convention 9.5.

- $I_1 = \left(\frac{18}{52}, \frac{19}{52}\right)$.
- $I_2 = \left(\frac{19}{52}, \frac{20}{52}\right)$ ($|I_1| = |I_2| = 17$).
- $I_3 = \left(\frac{22}{52}, \frac{23}{52}\right)$ ($|I_3| = 8$).

We need a finer classification of 3-students. We use Notation 9.4.

Claim 1: The only possible types of students are as follows:

(1) $(1, 1, 3)$ (we denote the number of such students by $y_{1,1,3}$).
(2) $(2, 2, 2)$ (we denote the number of such students by $y_{2,2,2}$).

Proof of Claim 1:
We show that some students are impossible.
A $(1, 2, 2)$-student has $< \frac{19}{52} + 2 \times \frac{20}{52} = \frac{59}{52}*$.
A $(1, 2, 3)$-student has $> \frac{18}{52} + \frac{19}{52} + \frac{22}{52} = \frac{59}{52}*$.

The result follows.

End of Proof of Claim 1

(We write down all equations that we know. This is overkill but it mirrors what we do in the general theorem.)

Since $|I_1| = |I_2| = 17$ we have $2y_{1,1,3} = 3y_{2,2,2} = 17$.
Since $|I_3| = 8$ we have $y_{1,1,3} = 8$.
Since $s_3 = 14$ we have $y_{1,1,3} + y_{2,2,2} = 14$.

One can check that this set of linear equations has no $\mathbb{N}$-solution. (Recall from Definition 5.5 that an $\mathbb{N}$-solution is a solution where every variable is in $\mathbb{N}$.) In fact, the first equation, $2y_{1,1,3} = 17$ already ensures no $\mathbb{N}$-solution. □

11.3. The Hard Buddy–Match Program

The proof that $f(59, 52) \le \frac{9}{26}$ went as follows:

(1) Determine what V is. If $V \ge 4$ then do not use this method.
(2) $V = 3$ so everyone is either a 2-student or a 3-student.
(3) Find out how many 2-students, 3-students, 2-shares, and 3-shares there are.
(4) Find the interval of 2-shares and the interval of 3-shares.
(5) Use a buddy–match sequence starting at the gap to find a gap that is properly within the 3-shares, splitting the 3-shares. This will lead to some intervals which have z shares, where we do not know z.
(6) Use a buddy–match sequence to find z.
(7) Use a buddy–match sequence to show that one of the intervals within the 3-shares is symmetric.
(8) Determine which types of students are possible.
(9) Use that information to form a set of linear equations that needs to have an $\mathbb{N}$-solution. (Recall from Definition 5.5 that an $\mathbb{N}$-solution is a solution where all of the variables are in $\mathbb{N}$.) If it doesn't then we get a contradiction. We will actually put a stricter condition on solutions. It will turn out that $s_3 = 2d$. Hence all of the variables are not just in $\mathbb{N}$ but are in $\{0, \ldots, 2d\}$. A solution must have all of its variables in $\{0, \ldots, 2d\}$. We call this *a 2d-solution*. When dealing with actual

values of m, s (e.g., $f(59, 52)$) this restriction is not useful since if any variable is $> 2d$ then some other one is < 0 since the sum of all the variables is $2d$. However, when proving general theorems, it will be useful to restrict solutions to $2d$-solutions.

We call this *the hard buddy–match (HBM) method*. It is harder than the easy buddy–match method since we need to look at types-of-students and linear equations.

Exercise 11.3. Prove each of the following using the HBM method:

(1) $f(25, 22) \leq \frac{23}{66}$.
(2) $f(33, 29) \leq \frac{10}{29}$.
(3) $f(34, 31) \leq \frac{32}{93}$.
(4) $f(38, 31) \leq \frac{11}{31}$.
(5) $f(43, 35) \leq \frac{5}{14}$.
(6) $f(41, 36) \leq \frac{37}{108}$.
(7) $f(43, 40) \leq \frac{41}{120}$.
(8) $f(49, 40) \leq \frac{57}{160}$.
(9) $f(45, 41) \leq \frac{14}{41}$.
(10) $f(49, 43) \leq \frac{44}{129}$.
(11) $f(55, 48) \leq \frac{83}{240}$.
(12) $f(59, 48) \leq \frac{103}{288}$.
(13) $f(52, 49) \leq \frac{50}{147}$.
(14) $f(60, 49) \leq \frac{5}{14}$.
(15) $f(57, 50) \leq \frac{17}{50}$.

We will now formalize the HBM-technique by doing the following:

(1) We will write a program VHBM(a, d, k, X) which uses the above template to, given a, d, k, X, output **Yes** if it proves that $f(3dk + a + d, 3dk + a) \leq \frac{dk+X}{3dk+a}$, and output NO otherwise.
(2) Once we have the program we will derive 12 corollaries that give *formulas* for upper bounds based on a, d. These corollaries find several

functions $X = X(a,d)$ (note that X does not depend on k) such that

$$(\forall k \geq 1) \left[f(3dk + a + d, 3dk + a) \leq \frac{dk + X}{3dk + a} \right].$$

11.4. The VHBM Algorithm

We will present the VHBM algorithm and, at the same time, prove that it works. Hence we will have both instructions for the algorithm and long passages that are then used to justify the instructions.

The algorithm VHBM below takes as input a, d, k, X and tries to verify that $f(3dk + a + d, 3dk + a) \leq \frac{dk+X}{3dk+a}$. There are some caveats:

- We require that $a \in \{1, \ldots, 3d\}$ since if $a \geq 3d + 1$ then one could take a larger value of k.
- We require that $k \geq 1$.
- We will first check if EBM already produced the bounds.

VHBM

- Input a, d, k, X.

Preprocessing Stage:

(1) If $a \notin \{1, \ldots, 3d\}$ then output BAD INPUT and stop.
(2) If $a = 2d$ then output FC($3dk + a + d, 3dk + a$). (This is Exercise 10.10.)
(3) If $X < \frac{a}{3}$ then output **DK** and stop. For the rest of the algorithm we assume $X \geq \frac{a}{3}$.
(4) If $X \geq \frac{a}{3}$ and $a \in \{2d+1, \ldots, 3d-1\}$ then output **Yes** and stop. (This is Theorem 10.11.1.) For the rest of the algorithm we assume $a \in \{1, \ldots, 2d-1\}$.
(5) If $X \geq \frac{a}{2}$ then output **Yes** and stop. (This is Theorem 10.11.2.) For the rest of the algorithm we assume $X < \frac{a}{2}$.
(6) If $X \geq \frac{a+d}{4}$ then output **Yes** and stop. (This is Theorem 10.11.3.) For the rest of the algorithm we assume $X < \frac{a+d}{4}$.

End of Preprocessing Stage

The following is not an algorithm, it is being used to set up the algorithm.

Getting More Information (Not an Algorithm)

Assume, by way of contradiction, that there is a $(3dk + a + d, 3dk + a)$-procedure with smallest piece $> \frac{dk+X}{3dk+a}$. We seek a contradiction.

Since $V = 3$ everyone has either 2 or 3 shares. As usual let s_2 (s_3) be the number of 2-students (3-students).

From the proof of Theorem 10.11 we know that:

(1) $s_2 = 3dk + a - 2d$, so there are $6dk + 2a - 4d$ 2-shares.
(2) $s_3 = 2d$, so there are $6d$ 3-shares.
(3) The following picture captures what we know:

$$\underset{\frac{dk+X}{3dk+a}}{(\quad 3s_3 \text{ 3-shs} \quad} \underset{\frac{dk+a+d-2X}{3dk+a}}{)[} \quad 0 \quad \underset{\frac{dk+d+X}{3dk+a}}{](} \quad 2s_2 \text{ 2-shs} \quad \underset{\frac{2dk+a-X}{3dk+a}}{)}$$

(The reader can check that the assumption $X \geq \frac{a}{3}$ ensures the interval of 2-shares and the interval of 3-shares do not intersect.)

We use a buddy–match sequence to find a useful empty interval.

$$M_0 = \left[\frac{dk+a+d-2X}{3dk+a}, \frac{dk+d+X}{3dk+a}\right],$$

$$B_0 = B(M_0) = \left[\frac{2dk+a-d-X}{3dk+a}, \frac{2dk-d+2X}{3dk+a}\right],$$

$$M_1 = M(B_0) = \left[\frac{dk+a+2d-2X}{3dk+a}, \frac{dk+2d+X}{3dk+a}\right],$$

$$B_1 = B(M_0) = \left[\frac{2dk+a-2d-X}{3dk+a}, \frac{2dk-2d+2X}{3dk+a}\right],$$

$$(\forall 0 \leq i \leq k - 1)\left[M_i = M(B_{i-1})\right.$$
$$= \left.\left[\frac{dk+a+(i+1)d-2X}{3dk+a}, \frac{dk+(i+1)d+X}{3dk+a}\right]\right],$$

$$(\forall 0 \leq i \leq k - 1)\left[B_i = B(M_i)\right.$$
$$= \left.\left[\frac{2dk+a-(i+1)d-X}{3dk+a}, \frac{2dk-(i+1)d+2X}{3dk+a}\right]\right].$$

We want that B_{k-2}, and hence $B_0, \ldots, B_{k-3}$, are contained in the 2-share region, so M can be applied to them. Note that

$$B_{k-2} = \left[\frac{dk+a+d-X}{3dk+a}, \frac{dk+d-X}{3dk+a}\right].$$

We need $\frac{dk+a+d-2X}{3dk+a} \leq \frac{dk+a+d-X}{3dk+a}$ which is clearly true.
Since M_0 is empty, so are all of the M_i's and B_i's. In particular

$$B_{k-1} = \left[\frac{dk+a-X}{3dk+a}, \frac{dk+2X}{3dk+a}\right]$$

is empty. We want that B_{k-1} is properly within the 3-shares. Hence we need

$$\frac{dk+2X}{3dk+a} < \frac{dk+a+d-2X}{3dk+a}$$

which we have since $X < \frac{a+d}{4}$. (This is the main place we use $X < \frac{a+d}{4}$. We also use it to simplify our table of X's later.)

Let the number of shares in $\left(\frac{dk+X}{3dk+a}, \frac{dk+a-X}{3dk+a}\right)$ be z. We discuss the number of shares each non-empty interval has. The reader may want to look at the next picture while reading this.

(1) $\left(\frac{dk+X}{3dk+a}, \frac{dk+a-X}{3dk+a}\right)$ and $\left(\frac{2dk+X}{3dk+a}, \frac{2dk+a-X}{3dk+a}\right)$ are buddies so they both have z shares.

(2) The $6d$ 3-shares are in $\left(\frac{dk+X}{3dk+a}, \frac{dk+a-X}{3dk+a}\right) \cup \left(\frac{dk+2X}{3dk+a}, \frac{dk+a+d-2X}{3dk+a}\right)$. Since the first interval has z, the second interval has $6d - z$.

(3) $\left(\frac{dk+2X}{3dk+a}, \frac{dk+a+d-2X}{3dk+a}\right)$ and $\left(\frac{2dk-d+2X}{3dk+a}, \frac{2dk+a-2X}{3dk+a}\right)$ are buddies so they both have $6d - z$ shares.

(4) The only non-empty interval not accounted for above is $\left(\frac{dk+d+X}{3dk+a}, \frac{2dk+a-d-X}{3dk+a}\right)$. Since there are $2(3dk+a+d) = 6dk+2a+2d$ shares total, this interval has

$$6dk + 2a + 2d - 2z - 2(6d - z) = 6dk - 10d + 2a \text{ shares.}$$

The following picture captures what we know (we do not know z):

The 3-shares:

$$(\underbrace{\quad z\text{ 3-shs}\quad}_{\frac{dk+X}{3dk+a}})[\underbrace{\quad 0\quad}_{\frac{dk+a-X}{3dk+a}}\ \underbrace{\quad}_{\frac{dk+2X}{3dk+a}}](\underbrace{\quad 6d-z\text{ 3-shs}\quad}_{\frac{dk+a+d-2X}{3dk+a}})[\underbrace{\quad 0\quad}_{\frac{dk+d+X}{3dk+a}}\ ]$$

The 2-shares:

$$(\underbrace{\quad 6dk-10d+2a\text{ 2-shs}\quad}_{\frac{dk+d+X}{3dk+a}})[\underbrace{\quad 0\quad}_{\frac{2dk+a-d-X}{3dk+a}}\ \underbrace{\quad}_{\frac{2dk-d+2X}{3dk+a}}]$$

$$(\underbrace{\quad 6d-z\text{ 2-shs}\quad}_{\frac{2dk-d+2X}{3dk+a}})[\underbrace{\quad 0\quad}_{\frac{2dk+a-2X}{3dk+a}}\ \underbrace{\quad}_{\frac{2dk+X}{3dk+a}}](\underbrace{\quad z\text{ 2-shs}\quad}_{\frac{2dk+a-X}{3dk+a}})$$

We use a buddy–match sequence to determine z.

$$M_0 = \left(\frac{dk+X}{3dk+a}, \frac{dk+d+X}{3dk+a}\right),$$

$$B_0 = B(M_0) = \left(\frac{2dk+a-d-X}{3dk+a}, \frac{2dk+a-X}{3dk+a}\right),$$

$$(\forall 0 \le i \le k-1)\left[M_i = M(B_{i-1})\right.$$
$$= \left.\left(\frac{dk+id+X}{3dk+a}, \frac{dk+(i+1)d+X}{3dk+a}\right)\right],$$

$$(\forall 0 \le i \le k-1)\left[B_i = B(M_i)\right.$$
$$= \left.\left(\frac{2dk+a-(i+1)d-X}{3dk+a}, \frac{2dk+a-id-X}{3dk+a}\right)\right].$$

We need this sequence to make sense. That is, we need that whenever we apply M to an interval, that interval is contained in the 2-share region. We proved this within the proof of Theorem 10.11.

The M_i's keep moving right. We do not want them to get to the end. Note that

$$M_{k-1} = \left(\frac{2dk-d+X}{3dk+a}, \frac{2dk+X}{3dk+a}\right).$$

So we want $2dk + X \le 2d + a - X$ which is true since $X \le \frac{a}{2}$.

The M_i's are disjoint and each has $3s_3$ shares

$$3ks_3 = |M_0 \cup \cdots \cup M_{k-1}| = \left|\left(\frac{dk+X}{3dk+a}, \frac{2dk+X}{3dk+a}\right)\right|.$$

But

$$\left|\left(\frac{dk+X}{3dk+a}, \frac{2dk+a-X}{3dk+a}\right)\right| = \text{all shares} = 2(3dk+a+d)$$
$$= 6dk+2a+2d.$$

So

$$z = \left|\left(\frac{2dk+X}{3dk+a}, \frac{2dk+a-X}{3dk+a}\right)\right| = 6dk+2a+2d-3ks_3$$
$$= 6dk+2a+2d-3k(2d)$$
$$= 2a+2d.$$

Great! We know z.
The following picture captures what we know:

The 3-shares:

$$\underset{\frac{dk+X}{3dk+a}}{(} \quad 2a+2d \text{ 3-shs} \quad \underset{\frac{dk+a-X}{3dk+a}}{)[} \quad 0 \quad \underset{\frac{dk+2X}{3dk+a}}{](} \quad 4d-2a \text{ 3-shs} \quad \underset{\frac{dk+a+d-2X}{3dk+a}}{)[} \quad 0 \quad \underset{\frac{dk+d+X}{3dk+a}}{]}$$

The 2-shares:

$$\underset{\frac{dk+d+X}{3dk+a}}{(} \quad 6dk-10d+2a \text{ 2-shs} \quad \underset{\frac{2dk+a-d-X}{3dk+a}}{)[} \quad 0 \quad \underset{\frac{2dk-d+2X}{3dk+a}}{]}$$

$$\underset{\frac{2dk-d+2X}{3dk+a}}{(} \quad 4d-2a \text{ 2-shs} \quad \underset{\frac{2dk+a-2X}{3dk+a}}{)[} \quad 0 \quad \underset{\frac{2dk+X}{3dk+a}}{](} \quad 2a+2d \text{ 2-shs} \quad \underset{\frac{2dk+a-X}{3dk+a}}{)}$$

The Hard Buddy–Match Method

We use a buddy–match sequence to find symmetry.

$$M_0 = \left(\frac{dk+X}{3dk+a}, \frac{dk+a-X}{3dk+a}\right),$$

$$B_0 = B(M_0) = \left(\frac{2dk+X}{3dk+a}, \frac{2dk+a-X}{3dk+a}\right),$$

$$M_1 = M(B_0) = \left(\frac{dk+d+X}{3dk+a}, \frac{dk+a+d-X}{3dk+a}\right),$$

$$B_1 = B(M_1) = \left(\frac{2dk-d+X}{3dk+a}, \frac{2dk+a-d-X}{3dk+a}\right),$$

$$(\forall 0 \le i \le k-1)\left[M_i = M(B_{i-1})\right.$$
$$= \left.\left(\frac{dk+id+X}{3dk+a}, \frac{dk+a+id-X}{3dk+a}\right)\right],$$

$$(\forall 0 \le i \le k-1)\left[B_i = B(M_i)\right.$$
$$= \left.\left(\frac{2dk-id+X}{3dk+a}, \frac{2dk+a-id-X}{3dk+a}\right)\right].$$

We leave it to the reader to prove that every time we apply M to an interval, that interval is contained in the 2-share region. We actually do not need to go all the way to $i = k-1$ to get to a symmetric interval. Where we go depends on the parity of k.

Case 1: k is even. Let $i = \frac{k}{2}$. Then

$$M_{k/2} = \left(\frac{dk+\frac{dk}{2}+X}{3dk+a}, \frac{dk+a+\frac{dk}{2}-X}{3dk+a}\right).$$

The sum of the endpoints is $\frac{3dk+a}{3dk+a} = 1$, so the midpoint is $\frac{1}{2}$. Hence $M_{k/2}$ is symmetric by buddying. In both cases M_0 is symmetric by the buddy–match sequence.

Case 2: k is odd. Let $i = \frac{k+1}{2}$. Then

$$M_{(k+1)/2} = \left(\frac{dk + \frac{d(k+1)}{2} + X}{3dk + a}, \frac{dk + a + \frac{d(k+1)}{2} - X}{3dk + a} \right).$$

The sum of the endpoints is $\frac{3dk+a+d}{3dk+a}$ so the midpoint is $\frac{1}{2} \times \frac{3dk+a+d}{3dk+a}$. Hence $M_{(k+1)/2}$ is symmetric by matching (see Section 10.2), and M_0 is symmetric by the buddy–match sequence.

The following picture captures what we know about the 3-shares:

$$\begin{array}{ccccc}
(\quad a+d \quad | \quad a+d \quad)[\quad 0 \quad](\quad 4d-2a \text{ 3-shs} \quad) \\
\frac{dk+X}{3dk+a} \quad \frac{dk+\frac{a}{2}}{3dk+a} \quad \frac{dk+a-X}{3dk+a} \quad \frac{dk+2X}{3dk+a} \quad \frac{dk+a+d-2X}{3dk+a}
\end{array}$$

We define the following intervals (and use Convention 9.5):

- $J_1 = \left(\frac{dk+X}{3dk+a}, \frac{dk+\frac{a}{2}}{3dk+a} \right)$,
- $J_2 = \left(\frac{dk+\frac{a}{2}}{3dk+a}, \frac{dk+a-X}{3dk+a} \right)$ $(|J_1| = |J_2| = a + d)$,
- $J_3 = \left(\frac{dk+2X}{3dk+a}, \frac{dk+a+d-2X}{3dk+u} \right)$ $(|J_3| = 4d - 2a)$.

For each type of student we determine the condition on X such that (1) the student does not get enough (that is, gets $< \frac{3dk+a+d}{3dk+a}$), and (2) the student gets too much (that is, gets $> \frac{3dk+a+d}{3dk+a}$). In either case, such a student does not exist.

(1, 1, 1)-students:

- Not enough: $3 \times \frac{dk+\frac{a}{2}}{3dk+a} \leq \frac{3dk+a+d}{3dk+a}$ which is $a \leq 2d$. Always true.
- Too much: $3 \times \frac{dk+X}{3dk+a} \geq \frac{3dk+a+d}{3dk+a}$. Impossible since $X < \frac{a+d}{4}$.

(1, 1, 2)-students:

- Not enough: $2 \times \frac{dk+\frac{a}{2}}{3dk+a} + \frac{dk+a-X}{3dk+a} \leq \frac{3dk+a+d}{3dk+a}$, which is $X \geq a - d$.
- Too much: $2 \times \frac{dk+X}{3dk+a} + \frac{dk+\frac{a}{2}}{3dk+a} \geq \frac{3dk+a+d}{3dk+a}$, which is $X \geq \frac{a+2d}{4}$. Impossible since $X < \frac{a+d}{4}$.

(1, 1, 3)-students:

- Not enough: $2 \times \frac{dk+\frac{a}{2}}{3dk+a} + \frac{dk+a+d-2X}{3dk+a} \leq \frac{3dk+a+d}{3dk+a}$, which is $X \geq \frac{a}{2}$. Impossible since $X < \frac{a}{2}$.

- Too much: $2 \times \frac{dk+X}{3dk+a} + \frac{dk+2X}{3dk+a} \geq \frac{3dk+a+d}{3dk+a}$, which is $X \geq \frac{a+d}{4}$. Impossible since $X < \frac{a+d}{4}$.

We stop here and leave the rest as an exercise.

Exercise 11.4. Show that the following table is correct: (The *Not Enough* column means that any student of that type gets $< \frac{3dk+a+d}{3dk+a}$. The *Too Much* column means that any student of that type gets $> \frac{3dk+a+d}{3dk+a}$.)

Student Type	Not Enough	Too Much
(1, 1, 1)	$0 = 0$	$0 \neq 0$
(1, 1, 2)	$X \geq a - d$	$0 \neq 0$
(1, 1, 3)	$0 \neq 0$	$0 \neq 0$
(1, 2, 2)	$X \geq \frac{3a-2d}{4}$	$0 \neq 0$
(1, 2, 3)	$0 \neq 0$	$X \geq \frac{a+2d}{6}$
(1, 3, 3)	$0 \neq 0$	$X \geq \frac{a+d}{5}$
(2, 2, 2)	$X \geq \frac{2a-d}{3}$	$0 \neq 0$
(2, 2, 3)	$0 \neq 0$	$X \geq \frac{d}{2}$
(2, 3, 3)	$0 \neq 0$	$X \geq \frac{a+2d}{8}$
(3, 3, 3)	$0 \neq 0$	$X \geq \frac{a+d}{6}$

Recall the following picture:

$$\underset{\frac{dk+X}{3dk+a}}{(\ a+d\ } \mid \underset{\frac{dk+\frac{a}{2}}{3dk+a}}{a+d\ )[} \underset{\frac{dk+a-X}{3dk+a}}{0\ ](} \underset{\frac{dk+2X}{3dk+a}}{} \underset{}{4d-2a\ \text{3-shs}} \underset{\frac{dk+a+d-2X}{3dk+a}}{)}$$

The picture leads us to the following equations:

Equations (1a) and (1b), respectively, based on $|J_1| = |J_2| = a + d$ and $|J_3| = 4d - 2a$:

$$3y_{1,1,1} + 2(y_{1,1,2} + y_{1,1,3}) + y_{1,2,2} + y_{1,2,3} + y_{1,3,3} = a + d, \quad (1a)$$

$$3y_{2,2,2} + 2(y_{1,2,2} + y_{2,2,3}) + y_{1,1,2} + y_{1,2,3} + y_{2,3,3} = a + d. \quad (1b)$$

Equation (2) based on $|J_3| = 4d - 2a$:

$$3y_{3,3,3} + 2(y_{1,3,3} + y_{2,3,3}) + y_{1,1,3} + y_{1,2,3} + y_{2,2,3} = 4d - 2a. \quad (2)$$

Equation (3) based on $s_3 = 2d$:

$$\sum_{1 \leq i \leq j \leq k \leq 3} y_{i,j,k} = 2d. \tag{3}$$

End of Getting More Information

We need a definition before resuming the algorithm.

Definition 11.5. A variable is *forbidden* (abbreviated *forb*) if a student of that type is forced to have either too much or too little muffin. We set such variables to 0. Other variables are called *permitted*.

We now resume the algorithm.

(1) Determine which of the $y_{i,j,k}$ variables are forbidden.
(2) In Eqs. (1)–(3) set the forbidden variables to 0. Denote the equations E.
(3) Determine if E has a $2d$-solution (all of the variables are in $\{0, \ldots, 2d\}$). If not then output **Yes**, if so then output **DK**.

11.5. Generating Formulas

We can use the Algorithm from Section 11.4 to generate theorems. We can find conditions on X that make many of the $y_{i,j,k}$ variables forbidden, and then condition on a, d such that the resulting set of equations has no $2d$-solution.

We will look at all ways to make 1, 2, 3, or 4 variables permitted. These will lead to formulas we express as corollaries.

From the table of condition on the $y_{i,j,k}$ variables, we have the following conditions:

- $y_{1,1,1}$ is always forbidden. We will not mention it again.
- $y_{1,1,3}$ is always permitted.
- $y_{1,2,3}$ forb $\implies$ $y_{1,3,3}$ forb $\implies$ $y_{2,3,3}$ forb $\implies$ $y_{2,2,2}$ forb. (For these $y_{i,j,k}$ the only way they are forbidden is if the student gets too much. Hence if another student gets more, they also get too much.)

- $y_{1,2,3}$ forb $\implies$ $y_{2,2,3}$ forb $\implies$ $y_{2,3,3}$ forb $\implies$ $y_{3,3,3}$ forb. (For these $y_{i,j,k}$ the only way they are forbidden is if the student gets too much. Hence if another student gets more, they also get too much.)
- $y_{2,2,2}$ forb $\implies$ $y_{1,2,2}$ forb $\implies$ $y_{1,1,2}$ forb. (For these $y_{i,j,k}$ the only way they are forbidden is if the student gets too little. Hence if another student gets less, they also get too little.)

From the above one can deduce the following:

(1) The only set of 1 variable such that one can make only the element of that set permitted is

- $\{y_{1,1,3}\}$.

(2) The only sets of 2 variables such that one can make only the elements of that set permitted are

- $\{y_{1,1,3}, y_{1,2,3}\}$,
- $\{y_{1,1,3}, y_{2,2,2}\}$.

(3) The only sets of 3 variables such that one can make only the elements of that set permitted are

- $\{y_{1,1,3}, y_{1,2,2}, y_{2,2,2}\}$,
- $\{y_{1,1,3}, y_{1,2,3}, y_{1,3,3}\}$,
- $\{y_{1,1,3}, y_{1,2,3}, y_{2,2,2}\}$,
- $\{y_{1,1,3}, y_{1,2,3}, y_{2,2,3}\}$.

(4) The only sets of 4 variables such that one can make only the elements of that set permitted are

- $\{y_{1,1,2}, y_{1,1,3}, y_{1,2,2}, y_{2,2,2}\}$,
- $\{y_{1,1,3}, y_{1,2,2}, y_{1,2,3}, y_{2,2,2}\}$,
- $\{y_{1,1,3}, y_{1,2,3}, y_{1,3,3}, y_{2,2,2}\}$,
- $\{y_{1,1,3}, y_{1,2,3}, y_{1,3,3}, y_{2,2,3}\}$,
- $\{y_{1,1,3}, y_{1,2,3}, y_{2,2,2}, y_{2,2,3}\}$.

We could list the only sets of 5 variables such that one can make only the elements of that set permitted. We do not, since this did not lead to any corollaries of interest. That is, the equations you get do not have an easy condition (e.g., $a \leq d$) such that they are unsolvable.

Notation 11.6. Let COND(X) be the following condition on X:

$$\frac{a}{3} \leq X < \min\left\{\frac{a}{2}, \frac{a+d}{4}\right\}.$$

In the next subsections we obtain 12 corollaries that will yield nice *formulas* to bound $f(3dk + a + d, 3dk + a)$. We only work out a few of them. For the rest we give the result but leave it to the reader to derive it.

We will sometimes state conditions on a, d that make the theorem work. The conditions will be labeled with a **, meaning they are useful, or with *Not Helpful*, which is what it sounds like, or with *Overshadowed* meaning a condition that could be helpful but for the fact that another condition implies it.

11.5.1. $y_{1,1,3}$ *Permitted, All Other Variables Forbidden*

To make all other variables forbidden:

- $y_{1,2,3}$ forbidden. The students get too much: $X \geq \frac{a+2d}{6}$.
- $y_{2,2,2}$ forbidden. The students get too little: $X \geq \frac{2a-d}{3}$.

If all variables except $y_{1,1,3}$ are set to 0, then the equations are

$$2y_{1,1,3} = a + d,$$
$$0 = a + d,$$
$$y_{1,1,3} = 4d - 2a,$$
$$y_{1,1,3} = 2d.$$

If these equations have a 2d-solution then $a + d = 0$ which is not possible. Hence we have the following corollary:

Corollary 11.7. *Let a, d, k be such that $d, k \geq 1, a \in \{1, \ldots, 2d\}$. Then*

$$f(3dk + a + d, 3dk + a) \leq \frac{dk + X}{3dk + a},$$

where $X = \max\left\{\frac{a+2d}{6}, \frac{2a-d}{3}\right\}$, if COND($X$) holds.

11.5.2. $y_{1,1,3}$, $y_{1,2,3}$ Permitted, All Other Variables Forbidden

Corollary 11.8. *Let a, d, k be such that $d, k \geq 1$, $a \in \{1, \ldots, 2d\}$. If $a \neq 1$ or $d \neq 1$, then*

$$f(3dk + a + d, 3dk + a) \leq \frac{dk + X}{3dk + a},$$

where $X = \max\left\{\frac{a+d}{5}, \frac{2a-d}{3}, \frac{d}{2}\right\}$, if COND(X) holds.

11.5.3. $y_{1,1,3}$, $y_{2,2,2}$ Permitted, All Other Variables Forbidden

Corollary 11.9. *Let a, d, k be such that $d, k \geq 1$, $a \in \{1, \ldots, 2d\}$. If $a \neq \frac{7d}{5}$, then*

$$f(3dk + a + d, 3dk + a) \leq \frac{dk + X}{3dk + a},$$

where $X = \max\{\frac{3a-2d}{4}, \frac{a+2d}{6}\}$, if COND(X) holds.

If $a = 10$ and $d = 7$ (note that $a \neq \frac{7d}{5}$), then $f(21k + 17, 21k + 10) \leq \frac{7k+4}{21k+10}$. Plug in $k = 2$ to get $f(59, 52) \leq \frac{18}{52}$, which is Theorem 11.2.

11.5.4. $y_{1,1,3}$, $y_{1,2,2}$, $y_{2,2,2}$ Permitted, All Other Variables Forbidden

To make all other variables forbidden:

- $y_{1,1,2}$ forbidden. The students get too little: $X \geq a - d$.
- $y_{1,2,3}$ forbidden. The students get too much: $X \geq \frac{a+2d}{6}$.

If all variables except $y_{1,1,3}$, $y_{1,2,2}$ and $y_{2,2,2}$ are set to 0 then the equations are:

$$2y_{1,1,3} + y_{1,2,2} = a + d,$$

$$2y_{1,2,2} + 3y_{2,2,2} = a + d,$$

$$y_{1,1,3} = 4d - 2a,$$

$$y_{1,1,3} + y_{1,2,2} + y_{2,2,2} = 2d.$$

This set of equations has solution:

(1) $y_{1,1,3} = 4d - 2a$
- $0 \leq 4d - 2a$, so $a \leq 2d$. Not helpful.
- $4d - 2a \leq 2d$, so $d \leq a$. Overshadowed.

(2) $y_{1,2,2} = 5a - 7d$
- $0 \leq 5a - 7d$, so $\frac{7d}{5} \leq a$**.
- $5a - 7d \leq 2d$, so $a \leq \frac{9d}{5}$. Overshadowed.

(3) $y_{2,2,2} = 5d - 3a$
- $0 \leq 5d - 3a$, so $a \leq \frac{5d}{3}$**.
- $5d - 3a \leq 2d$, so $d \leq a$. Overshadowed.

Corollary 11.10. Let a, d, k be such that $d, k \geq 1$, $a \in \{1, \ldots, 2d\}$. If $a < \frac{7d}{5}$ or $a > \frac{5d}{3}$, then

$$f(3dk + a + d, 3dk + a) \leq \frac{dk + X}{3dk + a},$$

where $X = \max\{a - d, \frac{a+2d}{6}\}$, if COND($X$) holds.

11.5.5. $y_{1,1,3}, y_{1,2,3}, y_{1,3,3}$ Permitted, All Other Variables Forbidden

Corollary 11.11. Let a, d, k be such that $d, k \geq 1$, $a \in \{1, \ldots, 2d\}$. If $a \neq 1$ or $d \neq 1$, then

$$f(3dk + a + d, 3dk + a) \leq \frac{dk + X}{3dk + a},$$

where $X = \max\{\frac{2a-d}{3}, \frac{d}{2}\}$, if COND($X$) holds.

11.5.6. $y_{1,1,3}, y_{1,2,3}, y_{2,2,2}$ Permitted, All Other Variables Forbidden

Corollary 11.12. Let a, d, k be such that $d, k \geq 1$, $a \in \{1, \ldots, 2d\}$. If $a < d$ or $a > \frac{7d}{5}$, then

$$f(3dk + a + d, 3dk + a) \leq \frac{dk + X}{3dk + a},$$

where $X = \max\{\frac{3a-2d}{4}, \frac{a+d}{5}, \frac{d}{2}\}$, if COND($X$) holds.

11.5.7. $y_{1,1,3}, y_{1,2,3}, y_{2,2,3}$ Permitted, All Other Variables Forbidden

Corollary 11.13. *Let a, d, k be such that $d, k \geq 1$, $a \in \{1, \ldots, 2d\}$. If $a \neq 1$ or $d \neq 1$, then*

$$f(3dk + a + d, 3dk + a) \leq \frac{dk + X}{3dk + a},$$

where $X = \max\{\frac{2a-d}{3}, \frac{a+d}{5}\}$, if COND(X) holds.

If $a = 4$ and $d = 3$, then $f(9k + 7, 9k + 4) \leq \frac{9k+5}{27k+12}$. Plug in $k = 2$ to get $f(25, 22) \leq \frac{23}{66}$.

11.5.8. $y_{1,1,2}, y_{1,1,3}, y_{1,2,2}, y_{2,2,2}$ Permitted, All Other Variables Forbidden

To make all other variables forbidden:

- $y_{1,2,3}$ forbidden. The students get too much: $X \geq \frac{a+2d}{6}$.

If all variables except $y_{1,1,2}, y_{1,1,3}, y_{1,2,2}, y_{2,2,2}$ are set to 0 then the equations are:

$$2y_{1,1,2} + 2y_{1,1,3} + y_{1,2,2} = a + d,$$

$$y_{1,1,2} + 2y_{1,2,2} + 3y_{2,2,2} = a + d,$$

$$y_{1,1,3} = 4d - 2a,$$

$$y_{1,1,2} + y_{1,1,3} + y_{1,2,2} + y_{2,2,2} = 2d.$$

These equations are not linearly independent hence they have an infinite number of solutions. We express the number of solutions by letting $y_{2,2,2} = w$ and expressing the other variables in terms of a, d, w. Since w is the number of students of type (2, 2, 2) and those must be 3-students, and there are $2d$ 3-students, we know that $0 \leq w \leq 2d$. We express these with a parameter w. Since $y_{2,2,2} = w$, $0 \leq w \leq 2d$.

(1) $y_{1,1,2} = 3a - 5d + w$

- $0 \leq 3a - 5d + 2d$ so $d \leq a$. Overshadowed.

- $3a - 5d \le 2d$ so $3a \le 7d$. Not useful.
- $0 \le 3a - 5d + w$ so $5d - 3a \le w$. So, $10d - 6a \le 2w$.
- $3a - 5d + w \le 2d$ so $w \le 7d - 3a$. So, $2w \le 14d - 6a$.

(2) $y_{1,1,3} = 4d - 2a$

- $0 \le 4d - 2a$, so $a \le 2d$. Not useful.
- $4d - 2a \le 2d$, so $d \le a$. Overshadowed.

(3) $y_{1,2,2} = 3d - a - 2w$

- $0 \le 3d - a$ so $a \le 3d$. Not useful.
- $3d - a - 4d \le 2d$ so $-a \le 3d$. Not useful.
- $0 \le 3d - a - 2w$ so $2w \le 3d - a$.
- $3d - a - 2w \le 2d$ so $d - a \le 2w$.

(4) $y_{2,2,2} = w$

Now look at the w's:
$10d - 6a \le 2w \le 3d - a$, so $\frac{7d}{5} \le a^{**}$.
$d - a \le 2w \le 14d - 6a$, so $a \le \frac{13d}{5}$. Not helpful.

Corollary 11.14. Let a, d, k be such that $d, k \ge 1$, $a \in \{1, \ldots, 2d\}$. If $a < \frac{7d}{5}$, then

$$f(3dk + a + d, 3dk + a) \le \frac{dk + X}{3dk + a},$$

where $X = \frac{a+2d}{6}$, if $\mathrm{COND}(X)$ holds.

11.5.9. $y_{1,1,3}$, $y_{1,2,2}$, $y_{1,2,3}$, $y_{2,2,2}$ Permitted, All Other Variables Forbidden

Corollary 11.15. Let a, d, k be such that $d, k \ge 1$, $a \in \{1, \ldots, 2d\}$. If $d > a$ or $a > \frac{7d}{3}$, then

$$f(3dk + a + d, 3dk + a) \le \frac{dk + X}{3dk + a},$$

where $X = \max\{a - d, \frac{a+d}{5}, \frac{d}{2}\}$, if $\mathrm{COND}(X)$ holds.

11.5.10. $y_{1,1,3}, y_{1,2,3}, y_{1,3,3}, y_{2,2,2}$ Permitted, All Other Variables Forbidden

Corollary 11.16. *Let a, d, k be such that $d, k \geq 1$, $a \in \{1, \ldots, 2d\}$. If $a < \frac{d}{2}$ or $a > \frac{7d}{5}$, then*

$$f(3dk + a + d, 3dk + a) \leq \frac{dk + X}{3dk + a},$$

where $X = \max\{\frac{3a-2d}{4}, \frac{d}{2}\}$, if COND(X) holds.

11.5.11. $y_{1,1,3}, y_{1,2,3}, y_{1,3,3}, y_{2,2,3}$ Permitted, All Other Variables Forbidden

Corollary 11.17. *Let a, d, k be such that $d, k \geq 1$, $a \in \{1, \ldots, 2d\}$. If $a > d$ or $a < \frac{d}{3}$, then*

$$f(3dk + a + d, 3dk + a) \leq \frac{dk + X}{3dk + a},$$

where $X = \max\{\frac{2a-d}{3}, \frac{a+2d}{8}\}$, if COND(X) holds.

11.5.12. $y_{1,1,3}, y_{1,2,3}, y_{2,2,2}, y_{2,2,3}$ Permitted, All Other Variables Forbidden

Corollary 11.18. *Let a, d, k be such that $d, k \geq 1$, $a \in \{1, \ldots, 2d\}$. If $a > \frac{7d}{5}$ or $a < d$, then*

$$f(3dk + a + d, 3dk + a) \leq \frac{dk + X}{3dk + a},$$

where $X = \max\{\frac{a+d}{5}, \frac{3a-2d}{4}\}$, if COND(X) holds.

Exercise 11.19. Use the corollaries above and the other techniques in this book to get formulas for $f(s+d, s)$ for small values of d. (We have a paper on the MUFFINS website that answers this.)

11.6. A Reasonable and a Bizarre Theorem

In all the corollaries presented in Section 11.5, we had a bound on $f(3dk + a + d, 3dk + a)$ of the form $\frac{dk+X}{3dk+a}$ where X did not depend on k. Hence the following theorem, due to Chatwin (2019), is not a surprise.

Theorem 11.20. (The Reasonable $X(a,d)$-Theorem). *For all a, d such that $a \in \{1, \ldots, 3d\}$, there exists $X = X(a, d)$ such that*

$$(\forall k \geq 1) \left[f(3dk + a + d, 3dk + a) \leq \frac{dk + X}{3dk + a} \right].$$

The proofs of the corollaries strongly used that there were only 2-shares and 3-shares, so buddy–match sequences were possible. This happens for $f(3dk + a + d, 3dk + a)$ when $k \geq 1$. What about when $k = 0$? In this case we do not have 2-shares and 3-shares, we have $(V-1)$-shares and V-shares for some $V \geq 3$. Hence, we cannot use buddy–match techniques. So there is no reason to expect that the formula above holds when $k = 0$. Except for one thing. It does. Chatwin also proved the following bizarre theorem.

Theorem 11.21. (The Unreasonable $X(a,d)$-Theorem). *For all a, d such that $a \in \{1, \ldots, 3d\}$, there exists $X = X(a, d)$ such that*

$$(\forall k \geq 0) \left[f(3dk + a + d, 3dk + a) \leq \frac{dk + X}{3dk + a} \right].$$

(For this conjecture we take, for all m, $f(m, 1) = f(m, 2) = \frac{1}{2}$ since that makes it work. This part is not unreasonable—we always assume that every muffin is cut into 2 pieces.)

Before we knew Theorem 11.21 was true we thought it was true and, by assuming it, got some real results. Here was the process:

(1) Given a, d we want to find the X such that

$$(\forall k \geq 1) \left[f(3dk + a + d, 3dk + a) \leq \frac{dk + X}{3dk + a} \right].$$

(2) Plug in $k = 0$ to get the expression $f(a + d, a) \leq \frac{X}{a}$.
(3) We actually know $f(a + d, a)$, so we can find a candidate for X.
(4) Given the candidate we verify it using buddy–match techniques.

(5) Once it's verified it doesn't matter that we got it by guesswork and a bizarre conjecture; it's true.

11.7. The Function HBM

We leave the formal definition of the HBM function to the reader. It is similar to the EBM function but with many more cases.

Theorem 11.22. *For every* $m \geq s$, $f(m, s) \leq \text{HBM}(m, s)$.

11.8. Program and Progress

Using the techniques presented so far we have the following attempt at an algorithm to find $f(m, s)$:

(1) Input(m, s).
(2) α is the min of
    {FC(m, s), Half(m, s), INT(m, s), MID(m, s), EBM(m, s), HBM(m, s)}.
(3) Run FINDPROC(m, s, α). If it outputs a procedure P then output α, else output **DK**.

There are 3520 pairs (m, s) we are considering (see Chapter 3). There were 78 pairs that neither FC nor Half nor INT nor MID nor EBM able to solve, but HBM was. Here are the full statistics so far. When we state that (say) for 329 cases $f(m, s) = \text{Half}(m, s)$, it is implicit that the prior techniques (in the case of Half its just FC) did not obtain the upper bound.

- For 2301 of them, $f(m, s) = \text{FC}(m, s)$. This is $\sim 65.37\%$.
- For 329 of them, $f(m, s) = \text{Half}(m, s)$. This is $\sim 9.35\%$.
- For 186 of them, $f(m, s) = \text{INT}(m, s)$. This is $\sim 5.28\%$.
- For 111 of them, $f(m, s) = \text{MID}(m, s)$. This is $\sim 3.15\%$.
- For 240 of them, $f(m, s) = \text{EBM}(m, s)$. This is $\sim 6.82\%$.
- For 89 of them, $f(m, s) = \text{HBM}(m, s)$. This is $\sim 2.53\%$.
- For 264 of them, none of FC, Half, INT, MID, EBM, or HBM suffices to find $f(m, s)$. This is $\sim 7.79\%$.

Chapter 12

The Gap and Train Methods

12.1. Recap and Goals for This Chapter

From Chapters 4, 6, and 8–11, we know that, for $m \geq s$,

$$f(m, s) \leq \min\{FC(m, s), \text{Half}(m, s), \text{INT}(m, s), \text{MID}(m, s),$$
$$\text{EBM}(m, s), \text{HBM}(m, s)\}.$$

Just for now we will refer to the min of those quantities as $\text{minf}(m, s)$. Is it the case that, for all $m \geq s$, $f(m, s) = \text{minf}(m, s)$? No. The counterexample with the smallest s is $f(47, 17)$; however it is more illustrative to show the following:

$$f(31, 19) \leq \frac{54}{133},$$
$$f(54, 47) \leq \frac{16}{47}.$$

The proof of these upper bounds uses a new technique, which we call *The Gap method*. We develop an algorithm $\text{Gap}(m, s)$ which, given m, s, finds an α such that $f(m, s) \leq \alpha$ can be proved by the Gap method.

Exercise 12.1. For $(m, s) = (47, 17), (31, 19), (54, 47)$:

(1) Compute $\alpha = \text{minf}(m, s)$.
(2) Compute $\text{FINDPROC}(m, s, \alpha)$. (You should get a **DK** which means that $\text{minf}(m, s)$ is unlikely to be $f(m, s)$.)

12.2. Thirty-One Muffins, Nineteen Students

Theorem 12.2. $f(31, 19) = \frac{54}{133}$.

Proof. We leave the proof that $f(31, 19) \geq \frac{54}{133}$ to the reader. Alternatively, the reader can run FINDPROC$(31, 19, \frac{54}{133})$.

We use denominator 266 to avoid fractional numerators. We prove $f(31, 19) \leq \frac{108}{266}$. Every student gets $\frac{31}{19} = \frac{434}{266}$.

We begin the proof that $f(31, 19) \leq \frac{54}{266}$ as if we were using the MID method.

Assume, by way of contradiction, that there is a (31, 19)-procedure with smallest piece $> \frac{108}{266}$. We leave it to the reader to show that there are fourteen 3-students, five 4-students, forty-two 3-shares, twenty 4-shares, and that the following picture captures what we know:

$$(\ 20 \ 4\text{-shs} \ )[\ 0 \ ](\ 22 \ 3\text{-shs} \ )[\ 0 \ ](\ 20 \ 3\text{-shs} \ )$$
$$\frac{108}{266} \qquad \frac{110}{266} \quad \frac{118}{266} \qquad \frac{148}{266} \quad \frac{156}{266} \qquad \frac{158}{266}$$

The following picture captures what we know about the 3-shares (we use Convention 9.5):

$$(\ 11 \ 3\text{-shs} \ | \ 11 \ 3\text{-shs} \ )[\ 0 \ ](\ 20 \ 3\text{-shs} \ )$$
$$\frac{118}{266} \qquad \frac{133}{266} \qquad \frac{148}{266} \quad \frac{156}{266} \qquad \frac{158}{266}$$

We define the following intervals:

- $I_1 = (\frac{118}{266}, \frac{133}{266})$.
- $I_2 = (\frac{133}{266}, \frac{148}{266})$ ($|I_1| = |I_2| = 11$).
- $I_3 = (\frac{156}{266}, \frac{158}{266})$ ($|I_3| = 20$).

We need a finer classification of 3-students. We use Notation 9.4.

Claim 1: The following are the only types of students who are possible:

(1) $(1, 2, 3)$ ($y_{1,2,3}$ denotes the number of such students).
(2) $(1, 3, 3)$ ($y_{1,3,3}$ denotes the number of such students).
(3) $(2, 2, 2)$ ($y_{2,2,2}$ denotes the number of such students).
(4) $(2, 2, 3)$ ($y_{2,2,3}$ denotes the number of such students).

Proof of Claim 1:
We establish that some students are impossible.

A $(1, 2, 2)$-student has $< \frac{133}{266} + 2 \times \frac{148}{266} = \frac{429}{266} < \frac{434}{266}$.
A $(1, 1, 3)$-student has $< 2 \times \frac{133}{266} + \frac{158}{266} = \frac{424}{266} < \frac{434}{266}$.
A $(2, 3, 3)$-student has $> \frac{133}{266} + 2 \times \frac{156}{266} = \frac{445}{266} > \frac{434}{266}$.
The result follows.

End of Proof of Claim 1

We set up equations just as with the MID technique:
Since $|I_1| = 11$:
$$y_{1,2,3} + y_{1,3,3} = 11.$$

Since $|I_2| = 11$:
$$y_{1,2,3} + 3y_{2,2,2} + 2y_{2,2,3} = 11.$$

Since $|I_3| = 20$:
$$y_{1,2,3} + 2y_{1,3,3} + y_{2,2,3} = 20.$$

Since $s_3 = 14$:
$$y_{1,2,3} + y_{1,3,3} + y_{2,2,2} + y_{2,2,3} = 14.$$

We need to show that this system has no $\mathbb{N}$-solutions.
Not so fast. It *does* have an $\mathbb{N}$-solution. In fact, it has four:

- $(y_{1,2,3}, y_{1,3,3}, y_{2,2,2}, y_{2,2,3}) = (2, 9, 3, 0)$;
- $(y_{1,2,3}, y_{1,3,3}, y_{2,2,2}, y_{2,2,3}) = (3, 8, 2, 1)$;
- $(y_{1,2,3}, y_{1,3,3}, y_{2,2,2}, y_{2,2,3}) = (4, 7, 1, 2)$;
- $(y_{1,2,3}, y_{1,3,3}, y_{2,2,2}, y_{2,2,3}) = (5, 6, 0, 3)$.

Hence the MID method fails to show $f(31, 19) \leq \frac{108}{266}$.

There was one hint that this proof would not work out: none of the proofs that students are impossible had a *.

Okay... now what? Look at the students who use I_1-shares. Assume Alice is a $(1, 2, 3)$-student. How big does the I_1-share have to be so that Alice can get enough?

(1) Since the I_2-share contributes $< \frac{148}{266}$ and the I_3-share contributes $< \frac{158}{266}$, together they contribute $< \frac{148}{266} + \frac{158}{266} = \frac{306}{266}$.
(2) Hence Alice's I_1-share must be $> \frac{434}{266} - \frac{306}{266} = \frac{128}{266}$.
(3) In short: A $(1, 2, 3)$-student has I_1-share $> \frac{128}{266}$.

Assume Bob is a (1, 3, 3)-student. How small does the I_1-share have to be to make sure that Bob doesn't get too much?

(1) The two I_3-shares contribute $> 2 \times \frac{156}{266} = \frac{312}{266}$.
(2) Hence Bob's I_1-share must be $< \frac{434}{266} - \frac{312}{266} = \frac{122}{266}$.
(3) In short: *A (1, 3, 3)-student has I_1-share $< \frac{122}{266}$*.

Since the only students who have I_1-shares are (1, 2, 3)-students and (1, 3, 3)-students we have:

There are no shares in $[\frac{122}{266}, \frac{128}{266}]$. And it gets better! By buddying: **There are no shares in $[\frac{138}{266}, \frac{144}{266}]$.**

The following picture captures what we know about the 3-shares (we do not know x or y):

$$(\ x \text{ 3-shs }\ )[\ 0 \ ](\ y \text{ 3-shs } \ | \ y \text{ 3-shs }\ )[\ 0 \ ](\ x \text{ 3-shs }\ )$$
$$\frac{118}{266} \qquad \frac{122}{266} \quad \frac{128}{266} \qquad \frac{133}{266} \qquad \frac{138}{266} \quad \frac{144}{266} \qquad \frac{148}{266}$$

$$[\ 0 \ ](\ 20 \text{ 3-shs }\ )$$
$$\frac{148}{266} \quad \frac{156}{266} \qquad \frac{158}{266}$$

We define new intervals (and use Convention 9.5).

- $I_1 = (\frac{118}{266}, \frac{122}{266})$.
- $I_2 = (\frac{128}{266}, \frac{133}{266})$.
- $I_3 = (\frac{133}{266}, \frac{138}{266})$ ($|I_2| = |I_3|$).
- $I_4 = (\frac{144}{266}, \frac{148}{266})$ ($|I_1| = |I_4|$).
- $I_5 = (\frac{156}{266}, \frac{158}{266})$ ($|I_5| = 20$).

(We also know that $|I_1| + |I_2| = |I_3| + |I_4| = 11$ but this is not needed.)

Claim 2: The following are the only types of students who are possible:

(1) (1, 5, 5) ($y_{1,5,5}$ denotes the number of such students).
(2) (2, 4, 5) ($y_{2,4,5}$ denotes the number of such students).
(3) (3, 4, 5) ($y_{3,4,5}$ denotes the number of such students).
(4) (4, 4, 4) ($y_{4,4,4}$ denotes the number of such students).

Proof of Claim 2:

We establish that some students are impossible.

A $(1, 4, 5)$-student has $< \frac{122}{266} + \frac{148}{266} + \frac{158}{266} = \frac{428}{266}$.
A $(3, 4, 4)$-student has $< \frac{138}{266} + 2 \times \frac{148}{266} = \frac{434}{266}*$.
A $(3, 3, 5)$-student has $< 2 \times \frac{138}{266} + \frac{158}{266} = \frac{434}{266}*$.
A $(2, 5, 5)$-student has $> \frac{128}{266} + 2 \times \frac{156}{266} = \frac{440}{266} > \frac{434}{266}$.
A $(4, 4, 5)$-student has $> 2 \times \frac{144}{266} + \frac{156}{266} = \frac{444}{266} > \frac{434}{266}$.
The result follows.

End of Proof of Claim 2

Since $|I_2| = |I_3|$:
$$y_{2,4,5} = y_{3,4,5}. \tag{1}$$

Since $|I_1| = |I_4|$:
$$y_{1,5,5} = y_{2,4,5} + y_{3,4,5} + 3y_{4,4,4}. \tag{2}$$

Since $s_3 = 14$:
$$y_{1,5,5} + y_{2,4,5} + y_{3,4,5} + y_{4,4,4} = 14. \tag{3}$$

We use Eq.(1) to eliminate $y_{3,4,5}$ from Eqs. (2) and (3) to get:
$$y_{1,5,5} = 2y_{2,4,5} + 3y_{4,4,4},$$
$$y_{1,5,5} = 14 - 2y_{2,4,5} - y_{4,4,4}.$$

We equate the two expressions for $y_{1,5,5}$ to get:
$$2y_{2,4,5} + 3y_{4,4,4} = 14 - 2y_{2,4,5} - y_{4,4,4}$$
$$2y_{2,4,5} + 4y_{4,4,4} = 7$$
$$y_{2,4,5} + 2y_{4,4,4} = \frac{7}{2},$$

which is a contradiction since this equation has no $\mathbb{N}$-solutions. $\square$

Exercise 12.3. Prove the following using the Gap method:

(1) $f(41, 19) \leq \frac{131}{304}$.
(2) $f(59, 22) \leq \frac{167}{374}$.

(3) $f(41, 23) \leq \frac{149}{368}$.
(4) $f(54, 25) \leq \frac{151}{350}$.
(5) $f(67, 25) \leq \frac{223}{500}$.
(6) $f(59, 26) \leq \frac{191}{442}$.
(7) $f(47, 29) \leq \frac{117}{290}$.
(8) $f(55, 31) \leq \frac{151}{372}$.
(9) $f(67, 31) \leq \frac{187}{434}$.
(10) $f(55, 34) \leq \frac{151}{374}$.

12.3. Fifty-Four Muffins, Forty-Seven Students

The proof of $f(31, 19) \leq \frac{54}{133}$ used buddying and the 3-shares. Most proofs using the Gap method just use buddying and one type of share.

We prove $f(54, 47) \leq \frac{16}{47}$ using buddying and matching (see Section 10.2). We will also use both the 2-shares and the 3-shares. There is one more nuance in the proof that we will point out when we get to it.

Theorem 12.4. $f(54, 47) = \frac{16}{47}$.

Proof. We leave the proof that $f(54, 47) \geq \frac{16}{47}$ to the reader. Alternatively, the reader can run FINDPROC(54, 47, $\frac{16}{47}$).

Since 47 divides 47 normally we would do this proof with denominator 47. However, if we did that then some of the other numbers involved would not be integers. We use denominator 94 to avoid that. We restate our theorem as $f(54, 47) \leq \frac{32}{94}$. Every student gets $\frac{54}{47} = \frac{108}{94}$.

Assume, by way of contradiction, that there is a (54, 47)-procedure with smallest piece $> \frac{32}{94}$. We leave it to the reader to show that there are thirty-three 2-students, fourteen 3-students, sixty-six 2-shares, forty-two 3-shares, and that the following picture captures what we know:

$$(\text{ 42 3-shs })[\ 0\ ](\text{ 66 2-shs })$$
$$\frac{32}{94} \qquad\qquad \frac{44}{94}\ \ \frac{46}{94} \qquad\qquad \frac{62}{94}$$

We would normally now buddy the gap to get another gap. Instead we will create a buddy–match sequence (see Definition 10.4) which will give us many gaps. The first gap we get is the usual buddy gap.

(1) $[\frac{44}{94}, \frac{46}{94}] = \emptyset$.
(2) $B[\frac{44}{94}, \frac{46}{94}] = [\frac{48}{94}, \frac{50}{94}]$ is empty. Since $[\frac{48}{94}, \frac{50}{94}]$ is contained in the 2-share region, we can apply M to it.
(3) $M[\frac{48}{94}, \frac{50}{94}] = [\frac{58}{94}, \frac{60}{94}]$ is empty.
(4) $B[\frac{58}{94}, \frac{60}{94}] = [\frac{34}{94}, \frac{36}{94}]$ is empty. Since $[\frac{34}{94}, \frac{36}{94}]$ overlaps the 3-shares region, we cannot apply M to it; we stop.

The following picture captures all we know about the 2-shares (we do not know x, y, or z).

$$(\ x \text{ 2-shares }\ )[\ 0\ ](\ y \text{ 2-shares }\ )[\ 0\ ](\ z \text{ 2-shares }\ )$$
$$\frac{46}{94} \quad\quad\quad\quad \frac{48}{94} \quad \frac{50}{94} \quad\quad\quad\quad \frac{58}{94} \quad \frac{60}{94} \quad\quad\quad\quad \frac{62}{94}$$

Are there symmetries? Yes:

(1) $(\frac{46}{94}, \frac{48}{94}) = (\frac{60}{94}, \frac{62}{94})$ by matching. So $x = z$.
(2) $(\frac{50}{94}, \frac{54}{94}) = (\frac{54}{94}, \frac{58}{94})$ by matching. So we want to break the interval $(\frac{50}{94}, \frac{58}{94})$ into two intervals $(\frac{50}{94}, \frac{54}{94})$ and $(\frac{54}{94}, \frac{58}{94})$ and then use Convention 9.5 so we do not need to worry about pieces of size $\frac{54}{94}$. But we can't! (This is the nuance we referred to above.) Let's say we did that and were wondering if 2-student Alice can have both shares in $(\frac{50}{94}, \frac{54}{94})$. Normally, we would say no since each piece is $< \frac{54}{94}$; hence Alice gets

$$< \frac{54}{94} + \frac{54}{94} = \frac{108}{94}.$$

Normally, this strict inequality is correct since at least one of the shares is the end of an open interval. But in this case both shares are the end of a closed interval. Alice really can't have two pieces in $(\frac{50}{94}, \frac{54}{94})$ but she can have two pieces in $[\frac{54}{94}, \frac{54}{94}]$. What to do? We make $[\frac{54}{94}, \frac{54}{94}]$ an interval all by itself. We call share sizes that need their own interval *troublesome*.

In the following picture, we are reusing the labels x, y, z. y and z do not mean what they meant in the previous picture. As usual, the following picture represents all we know about the 3-shares (we do not know x, y, or z).

(x 2-shs)[0](y 2-shs)[z 2-shs](y 2-shs)[0](x 2-shs)
$\frac{46}{94}$ $\frac{48}{94}$ $\frac{50}{94}$ $\frac{54}{94}$ $\frac{54}{94}$ $\frac{58}{94}$ $\frac{60}{94}$ $\frac{62}{94}$

Are there more symmetries? Yes:

(1) $|(\frac{46}{94}, \frac{47}{94})| = |(\frac{47}{94}, \frac{48}{94})|$ by buddying.
(2) $|(\frac{46}{94}, \frac{47}{94})| = |(\frac{61}{94}, \frac{62}{94})|$ by matching.
(3) $|(\frac{47}{94}, \frac{48}{94})| = |(\frac{60}{94}, \frac{61}{94})|$ by matching.
(4) $|(\frac{60}{94}, \frac{61}{94})| = |(\frac{61}{94}, \frac{62}{94})|$ by combining the above items.

In the following picture, we are reusing the labels x, y, z. x does not mean what it meant in the previous picture. As usual, the following picture represents all we know about the 2-shares (we do not know x, y, or z).

(x 2-shs | x 2-shs)[0](y 2-shs)[z 2-shs](y 3-shs)
$\frac{46}{94}$ $\frac{47}{94}$ $\frac{48}{94}$ $\frac{50}{94}$ $\frac{54}{94}$ $\frac{54}{94}$ $\frac{58}{94}$

[0](x 2-shs | x 2-shs)
$\frac{58}{94}$ $\frac{60}{94}$ $\frac{61}{94}$ $\frac{62}{94}$

Are there more symmetries? Yes:

(1) $|(\frac{32}{94}, \frac{33}{94})| = |(\frac{61}{94}, \frac{62}{94})|$ by buddying.
(2) $|(\frac{33}{94}, \frac{34}{94})| = |(\frac{60}{94}, \frac{61}{94})|$ by buddying.
(3) $|(\frac{32}{94}, \frac{33}{94})| = |(\frac{33}{94}, \frac{34}{94})|$ by the above and $(\frac{60}{94}, \frac{61}{94}) = (\frac{61}{94}, \frac{62}{94})$.
(4) All of the above intervals in the last three items have x shares since both $(\frac{46}{94}, \frac{47}{94})$ and $(\frac{47}{94}, \frac{48}{94})$ have x shares. This is easy and we leave it to the reader.
(5) $[\frac{54}{94}, \frac{54}{94}] = [\frac{40}{94}, \frac{40}{94}]$ by buddying. Even though $\frac{40}{94}$ is not a troublesome share size (recall that $\frac{54}{94}$ is troublesome), for symmetry we also give $\frac{40}{94}$ its own interval.
(6) $|(\frac{36}{94}, \frac{40}{94})| = |(\frac{54}{94}, \frac{58}{94})|$ by buddying.
(7) $|(\frac{40}{94}, \frac{44}{94})| = |(\frac{50}{94}, \frac{54}{94})|$ by buddying.
(8) $|(\frac{36}{94}, \frac{40}{94})| = |(\frac{40}{94}, \frac{44}{94})|$ by the above and $|(\frac{50}{94}, \frac{54}{94})| = |(\frac{54}{94}, \frac{58}{94})|$.

The Gap and Train Methods

The following picture captures everything we know about the 2-shares and the 3-shares.

The 3-shares:

$$(\; x \text{ 3-shs} \; | \; x \text{ 3-shs} \;)[\; 0 \;](\; y \text{ 3-shs} \;)[\; z \text{ 3-shs} \;](\; y \text{ 3-shs} \;)$$
$$\tfrac{32}{94} \qquad \tfrac{33}{94} \qquad\qquad \tfrac{34}{94} \quad \tfrac{36}{94} \qquad\qquad \tfrac{40}{94} \qquad\qquad \tfrac{40}{94} \qquad\qquad \tfrac{44}{94}$$

The 2-shares:

$$(\; x \text{ 2-shs} \; | \; x \text{ 2-shs} \;)[\; 0 \;](\; y \text{ 2-shs} \;)[\; z \text{ 2-shs} \;](\; y \text{ 2-shs} \;)$$
$$\tfrac{46}{94} \qquad \tfrac{47}{94} \qquad\qquad \tfrac{48}{94} \quad \tfrac{50}{94} \qquad\qquad \tfrac{54}{94} \qquad\qquad \tfrac{54}{94} \qquad\qquad \tfrac{58}{94}$$

$$[\; 0 \;](\; x \text{ 2-shs} \; | \; x \text{ 2-shs} \;)$$
$$\tfrac{58}{94} \quad \tfrac{60}{94} \qquad\qquad \tfrac{61}{94} \qquad\qquad \tfrac{62}{94}$$

We define two types of intervals. The J-intervals are of 3-shares, the I-intervals are of 2-shares. Our explanations of why certain intervals are equal are either *mat* for matching, *bud* for buddying, or *trans* for transitive closure of the other equalities.

- $J_1 = (\frac{32}{94}, \frac{33}{94})$.
- $J_2 = (\frac{33}{94}, \frac{34}{94})$ ($|J_1| = |J_2|$ by trans).
- $J_3 = (\frac{36}{94}, \frac{40}{94})$.
- $J_4 = [\frac{40}{94}, \frac{40}{94}]$.
- $J_5 = (\frac{40}{94}, \frac{44}{94})$ ($|J_3| = |J_5|$ by trans).
- $I_1 = (\frac{46}{94}, \frac{47}{94})$.
- $I_2 = (\frac{47}{94}, \frac{48}{94})$ ($|I_1| = |I_2|$ by bud).
- $I_3 = (\frac{50}{94}, \frac{54}{94})$ ($|J_5| = |I_3|$ by bud).
- $I_4 = [\frac{54}{94}, \frac{54}{94}]$ ($|J_4| = |I_4|$ by bud).
- $I_5 = (\frac{54}{94}, \frac{58}{94})$ ($|J_3| = |I_5|$ by bud, $|I_3| = |I_5|$ by mat).
- $I_6 = (\frac{60}{94}, \frac{61}{94})$ ($|J_2| = |I_6|$ by bud, $|I_2| = |I_6|$ by mat).
- $I_7 = (\frac{61}{94}, \frac{62}{94})$ ($|J_1| = |I_7|$ by bud, $|I_1| = |I_7|$ by mat).

Since $|J_1| = |I_7|$ and $|I_1| = |I_7|$, we have $|I_1| = |J_1|$. Since $|I_1| = |I_2|$ and $|J_1| = |J_2|$, we have

$$|J_1| = |J_2| = |I_1| = |I_2|.$$

More equations can be obtained; however, we will not need them.

The astute reader may have noticed that we have more troublesome share sizes. Let's say we were wondering if there can be a $(1, 6)$-student Alice, so Alice would have one share in I_1 and one share in I_6. Normally, we would say this is not possible since she would have

$$< \frac{47}{94} + \frac{61}{94} = \frac{108}{94}.$$

Normally, this strict inequality is correct since at least one of the shares is the end of an open interval. But there can be shares of sizes $\frac{47}{94}$ and $\frac{61}{94}$. We should declare both $\frac{47}{94}$ and $\frac{61}{94}$ troublesome and make them each their own interval, plus their buddies (though $\frac{47}{94}$ is its own buddy, so no need there). *We are not going to do this.* Hence our proof is technically incorrect. We leave it to the reader to modify the proof to make it correct. As it stands, the proof still illustrates the points we want to make, that is, if $V = 3$ then a gap proof can use a buddy–match sequence and symmetries.

Claim 1: The following are the only types of students who are possible:

(1) $(1, 1, 5)$ ($y_{1,1,5}$ will denote the number of these).
(2) $(1, 2, 5)$ ($y_{1,2,5}$ will denote the number of these).
(3) $(1, 3, 3)$ ($y_{1,3,3}$ will denote the number of these).
(4) $(2, 2, 5)$ ($y_{2,2,5}$ will denote the number of these).
(5) $(2, 3, 3)$ ($y_{2,3,3}$ will denote the number of these).

Proof of Claim 1:
We establish that some students are impossible.
  A $(2, 2, 4)$-student has $< 2 \times \frac{34}{94} + \frac{40}{94} = \frac{108}{94}*$.
  A $(3, 3, 3)$-student has $> 3 \times \frac{36}{94} = \frac{108}{94}*$.
  A $(1, 3, 4)$-student has $> \frac{32}{94} + \frac{36}{94} + \frac{40}{94} = \frac{108}{94}*$.

End of Proof of Claim 1

Claim 2: The following are the only types of students who are possible:

(1) (1, 7) ($z_{1,7}$ will denote the number of these).
(2) (2, 6) ($z_{2,6}$ will denote the number of these).
(3) (3, 5) ($z_{3,5}$ will denote the number of these).

Proof of Claim 2: We establish that some students are impossible.

A (1, 6)-student has $< \frac{47}{94} + \frac{61}{94} = \frac{108}{94}*$.
A (2, 5)-student has $< \frac{48}{94} + \frac{58}{94} = \frac{106}{94} < \frac{108}{94}$.
A (2, 7)-student has $> \frac{47}{94} + \frac{61}{94} = \frac{108}{94}*$.
A (3, 6)-student has $> \frac{50}{94} + \frac{60}{94} = \frac{110}{94} > \frac{108}{94}$.

By Claim 1 there are no shares in J_4. Since I_4 buddies J_4 there are no shares in I_4. Hence there are no students of type (3, 4), (4, 4), (4, 5), (4, 6), or (4, 7).

The result follows.

End of Proof of Claim 2

Since $|I_1| = |I_2|$:
$$z_{1,7} = z_{2,6}. \tag{1}$$

Since $|J_1| = |J_2| = |I_1| = |I_2|$:
$$2y_{1,1,5} + y_{1,2,6} + y_{1,3,3} = 2y_{2,2,5} + y_{2,3,3} = z_{1,7}. \tag{2}$$

Since $|J_5| = |I_3|$:
$$y_{1,1,5} + y_{1,2,5} + y_{2,2,5} = z_{3,5}. \tag{3}$$

Since $|J_3| = |I_5|$:
$$2y_{1,3,3} + 2y_{2,3,3} = z_{3,5}. \tag{4}$$

Since $s_2 = 33$:
$$z_{1,7} + z_{2,6} + z_{3,5} = 33. \tag{5}$$

Since $s_3 = 14$:
$$y_{1,1,5} + y_{1,2,5} + y_{1,3,3} + y_{2,2,5} + y_{2,3,3} = 14. \tag{6}$$

We rewrite Eq. (6) as
$$(y_{1,1,5} + y_{1,2,5} + y_{2,2,5}) + \frac{1}{2} \times (2y_{1,3,3} + 2y_{2,3,3}) = 14. \tag{7}$$

Using Eqs. (3) and (4) we rewrite Eq. (7) as

$$z_{3,5} + \frac{1}{2}z_{3,5} = 14$$

$$3z_{3,5} = 28.$$

Since 28 is not a multiple of 3, there is no ℕ-solutions. □

At the end of the proof of Theorem 12.4 we needed to show that a set of linear equations did not have a ℕ-solution. In this case, we were lucky that we could prove there was no ℕ-solution by hand. For other problems we are (1) not so lucky, in that hand calculations would be tedious, but (2) lucky that we live in the modern computer age. Indeed, we use a linear equation solver.

12.4. The VGap Program

The following pseudocode summarizes the VGap method:

VGap

(1) Input(m, s, α).
(2) Find intervals and gaps as in the VMID method. Hence the interval that contains $\frac{1}{2}$ will be split into two intervals that buddy each other. If $V = 3$ then (1) use a buddy–match sequence to find intervals and gaps, (2) split the interval that has $\frac{m}{2s}$, and (3) make $\frac{m}{2s}$ its own interval.
(3) Iterate the following process until no new gaps are formed.
   (a) Note which intervals have the same number of shares via buddying. If $V = 3$ then note which intervals of 2-shares have the same number of shares via matching and make the troublesome shares, and their buddies and matches, into separate intervals.
   (b) Find the possible students.
   (c) Use them to try to create new gaps. If such are found then use buddying (and a buddy–match sequence if $V = 3$) to find more intervals and gaps.
(4) (Iteration will produce no new gaps.) Assign to each possible type of student a variable. Set up the equations based on what you know about

symmetry, how much is in some intervals, s_V and s_{V-1}. If this system has no $\mathbb{N}$-solution then you are done, $f(m,s) \le \alpha$, so output **Yes**. Else output **DK**.

12.5. The Gap Program

In the prior chapters we had a program for (say) Half that *produced* an upper bound. So far we have VGap which *verifies* upper bounds but does not produce them. In order to produce an upper bound, we need to do a binary-search-like algorithm with VGap.

Gap

(1) Input(m, s).
(2) Look at all the rationals between $\frac{1}{3}$ and $\frac{1}{2}$ that have denominators of the form bs where $1 \le b \le ms$. (Empirical evidence suggests that the denominator is always a multiple of s, that is, $\le ms^2$. It is usually much lower.) Sort them to produce $\alpha_1 < \alpha_2 < \cdots < \alpha_n$. Add in $\alpha_0 = \frac{1}{3}$ if it is not there already.
By using VGap(m, s, α_i) do a binary search to find the smallest α_i such that VGap$(m, s, \alpha_i) = $ **Yes**. Output α_i.

Note 12.5. In practice, we will know much better upper bounds that $\frac{1}{2}$ having ran FC(m, s), Half(m, s), etc. Hence the binary search will be on less numbers than portrayed here.

Theorem 12.6. *For all $m \ge s$, $f(m, s) \le $ Gap(m, s).*

12.6. Program and Progress

Using the techniques presented so far we have the following attempt at an algorithm to find $f(m, s)$:

(1) Input(m, s).
(2) α is the min of
{FC(m, s), Half(m, s), INT(m, s),
MID(m, s), EBM(m, s), HBM(m, s), Gap(m, s)}.

(3) Run FINDPROC(m, s, α). If it outputs a procedure then output α, else output **DK**.

12.7. The Train Method

Before the Gap method we had solved 3246 of the 3520 cases we are considering (see Chapter 3), leaving 274 unsolved. The Gap method solved 261 of them, leaving 13 unsolved. Hence we need a new method. We found one! The Train method. We omit it from this book since it is somewhat complicated; however, it is on the MUFFIN website.

The Train method did indeed solve the remaining 13 cases.

12.8. Program

Using the techniques presented so far (and the Train method), we have the following attempt at an algorithm to find $f(m, s)$. Since it is the last algorithm using our methods, we call it OURULTIMATE.

OURULTIMATE

(1) Input(m, s).
(2) α is the min of

{FC(m, s), Half(m, s), INT(m, s), MID(m, s), EBM(m, s), HBM(m, s)}

(3) Run FINDPROC(m, s, α). If it outputs a procedure P then output α.
(4) Run Gap(m, s) except that we can do the binary search with α as the upper bound, not $\frac{1}{2}$. If it outputs α then by the Gap Program, α is the answer, so output α.
(5) Run Train(m, s) except that we can do the binary search with α as the upper bound, not $\frac{1}{2}$. If it outputs α then by the Train Program, α is the answer, so output α. (Train(m, s) will also use some of the information that Gap(m, s) found in its failed attempt to prove the upper bound.)

There are 3520 pairs (m, s) we are considering (see Chapter 3). There were 261 pairs that neither FC nor Half nor INT nor MID nor EBM nor HBM were able to solve, but Gap was. There were 14 pairs where none of the above could solve them, nor could Gap, but Train could. Here are the full statistics. When we state that (say) for 329 cases $f(m, s) = \text{Half}(m, s)$, it is implicit that the prior techniques (in the case of Half its just FC) did not obtain the upper bound.

- For 2301 of them, $f(m, s) = \text{FC}(m, s)$. This is $\sim 65.37\%$.
- For 329 of them, $f(m, s) = \text{Half}(m, s)$. This is $\sim 9.35\%$.
- For 186 of them, $f(m, s) = \text{INT}(m, s)$. This is $\sim 5.28\%$.
- For 111 of them, $f(m, s) = \text{MID}(m, s)$. This is $\sim 3.15\%$.
- For 240 of them, $f(m, s) = \text{EBM}(m, s)$. This is $\sim 6.82\%$.
- For 89 of them, $f(m, s) = \text{HBM}(m, s)$. This is $\sim 2.53\%$.
- For 250 of them, $f(m, s) = \text{Gap}(m, s)$ This is $\sim 7.10\%$.
- For 13 of them, $f(m, s) = \text{Train}(m, s)$. This is $\sim 0.4\%$.
- All 3520 are solvable by one of the methods above.

We have solved all of the problems we were considering. Yeah! So... are we done?

The next chapter presents a method, due to Scott Huddleston, which, given (m, s), finds an (m, s)-procedure very quickly. In all cases we have tested it gives the optimal one. Later, Chatwin (2019) rediscovered Scott's algorithm and proved it correct. Hence we actually know all of the answers. So, do our techniques always lead to the correct answer? Alas no. The following were discovered by Scott's algorithm; however, our techniques cannot prove the upper bounds:

(1) $f(205, 178) = \frac{214}{623}$.
(2) $f(226, 135) = \frac{388}{945}$.
(3) $f(233, 141) = \frac{691}{1692}$.

We suspect that we could extend our methods to prove these three. However, this would get more complicated and actually approach being the harder methods of Scott that we have been trying to avoid. Hence, for our own techniques, we stop here.

Chapter 13

Scott Huddleston's Method*

13.1. Recap

In Section 12.8, we used *all* of our methods to produce a program that computes $f(m, s)$ for all 3519 pairs we targeted in Chapter 3. This should be cause for celebration! But (1) the program runs slowly for some values of m, s, and (2) there are cases where our methods do not suffice (see the end of Chapter 12).

We discuss a method that was discovered independently by Scott Huddleston and Richard Chatwin. We call this *Scott's Algorithm* since Scott was the first to contact us about it. Richard Chatwin proved that Scott's algorithm always finds the answer [Chatwin (2019)], though we omit the proof. Scott's algorithm is extremely fast in both theory and practice. Formally, Scott's algorithm shows that finding $f(m, s)$ and an (m, s)-procedure that achieves this bound can be done in time $O(m^2)$. The analysis may not be tight—it is possible that the algorithm always runs in $O(m \log n)$ time. In any case, in practice, it is *always* fast.

We give two examples of Scott's algorithm. These examples will help to establish notation and concepts. We then present Scott's algorithm.

13.2. Five Muffins, Three Students

We assume $f(5, 3) > \frac{1}{3}$. By Theorem 2.6, every muffin is cut into 2 pieces, so there are 10 shares. We leave it to the reader to show that there are two

3-students, one 4-student, six 3-shares, and four 4-shares. So far this is all standard.

We introduce several new ideas that we use throughout this section.

New Idea One: Generalize the Problem. We first restate the (5,3)-problem:

(1) We have 5 muffins that are worth 1 each and cut into 2 pieces. We denote this as $(5, 1, 2)$. In the future we will have muffins that have values other than 1.
(2) We have one 4-student who needs $\frac{5}{3}$ via 4 shares. We denote this as $(1, \frac{5}{3}, 4)$.
(3) We have two 3-students who need $\frac{5}{3}$ via 3 shares. We denote this as $(2, \frac{5}{3}, 3)$.

We denote this problem as

$$\text{Scott}\left[(5, 1, 2), \left(1, \frac{5}{3}, 4\right), \left(2, \frac{5}{3}, 3\right)\right].$$

We call it SC(5, 3)-0. We will soon recast it as a problem about finding weights on edges in a graph. We will still call this recast problem SC(5, 3)-0.

This is an example of a *Scott-Muffin problem*. We now give the formal definition and conventions.

Definition 13.1. A *Scott-Muffin problem* is a 3-tuple of 3-tuples:

$$(n_m, v_m, p_m),$$
$$(n_{s_1}, v_{s_1}, p_{s_1}), \qquad (13.1)$$
$$(n_{s_2}, v_{s_2}, p_{s_2})$$

with the following meaning:

(1) All three tuples are in $\mathbb{N} \times \mathbb{Q} \times \mathbb{N}$. All 9 numbers are ≥ 0.
(2) The first tuple (n_m, v_m, p_m) means that there are n_m muffins, each with value v_m, and each cut into p_m pieces. Later p_m will be the degree of a muffin-vertex in a graph. All three of these numbers are > 0.
(3) Both the second and third tuples represent students.

(a) The second tuple $(n_{s_1}, v_{s_1}, p_{s_1})$ means that there are n_{s_1} students (these are not all of the students) who want muffins of value v_{s_1}, in p_{s_1} shares. These are called *the major students* (we'll see why in point (c)). Later p_{s_1} will be the degree of a student-vertex in a graph. All three of these numbers are > 0.

(b) The third tuple $(n_{s_2}, v_{s_2}, p_{s_2})$ means there are n_{s_2} students who want muffins of value v_{s_2}, in p_{s_2} shares. These are called *the minor students* (we'll see why in point (c)). Later p_{s_2} will be the degree of a student-vertex in a graph. If all three numbers are 0 then we leave it off and in this case the Scott-Muffin problem only has two tuples—a muffin tuple and the major students.

(c) Which student-tuple is major and which is minor is determined as follows: the tuple with the larger ratio $\frac{\text{degree}}{\text{value}}$ are the major muffins. In other words, $\frac{p_{s_1}}{v_{s_1}} > \frac{p_{s_2}}{v_{s_2}}$.

(4) Be forewarned: you are used to thinking of pieces of muffins being given to students. We will often invert that and think of students giving pieces to muffins. The graphs we use will be undirected so either mentality is fine.

The Scott-Muffin problem is important since we will be taking a standard muffin problem and transforming it into a smaller Scott-Muffin problem, and then (possibly) again into an even smaller Scott-Muffin problem, until we get to a certain type of Scott-Muffin problem that is easy to solve optimally. We will then use that to solve all the problems in the sequence (conjecturally) optimally. So solving Scott-Muffin problems is an example of that old chestnut: *it's sometimes easier to solve a harder problem.*

New Idea Two: Represent the Problem as a Graph. Since the 4-student only uses 4 shares and there are 5 muffins, there must be a muffin that is shared *among only the 3-students*. Since each muffin is cut in two pieces, there will be two 3-students who share a muffin. We represent this in Graph 13.1 where the massive magenta[1] vertices are muffins and the small cyan vertices are students.

---

[1] Depending on the medium you are reasoning this in you may or may not see the colors.

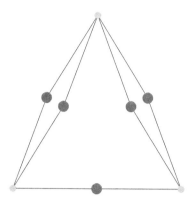

Graph 13.1. Five Muffins, Three Students, SC(5, 3)-0.

We will present many more graphs (actually multigraphs) where (1) vertices are either students or muffins, and (2) a muffin vertex is connected to a student vertex if that student gets a piece of that muffin. We state the conventions for such graphs.

Convention 13.2. In all of our graphs, the following conditions hold:

(1) Muffins are Massive Magenta (reddish) colored dots. (M for Muffin, Massive, and Magenta).
(2) Students are Small Cyan (bluish) colored dots (S for Student, small, and (sort of) Cyan).
(3) A muffin and a student are connected if a student has a piece of that muffin. Since muffins can only be connected to students and vice versa, students and muffins are the two parts of a bipartite graph. We do not draw the graphs as bipartite since such a drawing would be a mess.
(4) We associate to a Scott-Muffin problem

$$\begin{aligned}(n_m, v_m, p_m),\\ (n_{s_1}, v_{s_1}, p_{s_1}),\\ (n_{s_2}, v_{s_2}, p_{s_2})\end{aligned} \quad (13.2)$$

a graph. This graph is *not unique*. That is, there may be more than one graph that represents the problem. This will end up not mattering since the graphs are visual aids and not used in the actual algorithm. The graph will have n_m muffin-vertices of degree p_m, n_{s_1} student-vertices of degree p_{s_1}, and n_{s_2} student-vertices of degree p_{s_2} (we leave out for now how to determine the edges). The problem is to assign nonnegative weights to the edges such that every muffin-vertex has weighted degree v_m, that every major-student-vertex has weighted degree v_{s_1}, and that every minor-student-vertex has weighted degree v_{s_2}. It is easy to see how these weights can be used to obtain a solution to the Scott-Muffin problem.

(5) Note that (1) all of the muffin-nodes are of degree p_m, (2) all of the major-student-nodes are of degree p_{s_1}, and (3) all of the minor-student-nodes are of degree p_{s_2}.

(6) Note that the graph itself does not specify the entire Scott-Muffin problem. We often say things like *this graph captures some of the Scott-Muffin problem*.

(7) The muffin-vertices for a standard muffin problem will have degree 2 since each muffin is cut into exactly 2 pieces. For a Scott-Muffin problem where the muffins may have values other than 1 and may be cut into more than 2 pieces, the muffin-vertices may have higher degree.

(8) The weights on the edges represent the size of the piece that the muffin gave to the student. To re-iterate: we will often invert that and think of a student giving pieces to a muffin.

Since there are two 3-students who share a muffin, and one 4-student, Graph 13.1 captures some of what we know.

Note that a (5, 3)-procedure is a way to assign nonnegative weights to the edges of Graph 13.1 such that:

- The weighted degree of each muffin vertex is 1.
- The weighted degree of each student vertex is $\frac{5}{3}$.

We call the problem of finding such weights SC(5, 3)-0.

New Idea Three: Transform the Problem into a Smaller One—Clusters Are Students. We need a notation for a certain part of the graph.

Definition 13.3. Let $L \geq 0$. An *L-cluster* is a sequence of length $2L + 1$ of the form student–muffin–$\cdots$–student that has L minor students, together with all the other muffins attached to the students. The muffins in the student–muffin–$\cdots$–student sequence are called *internal muffins* whereas the muffins that are not in that sequence but are attached to the students are called *external muffins*. The muffins in the sequence might have other students attached to them but those students are not part of the cluster.

Graph 13.1 has a 1-cluster consisting of the 2 students and 1 internal muffin (at the bottom) *together with* the four external muffins that are adjacent to the students. Here is the big new idea:

We will transform the problem by regarding this 1-cluster as being a student.

The internal and external muffins are part of the cluster. The 2 students in the cluster need $2 \times \frac{5}{3} = \frac{10}{3}$. There is 1 internal muffin and there are 4 external muffins so the cluster has 5 muffins. We are now going to view the students as having muffin pieces to give to the muffins. Hence the cluster can be viewed as *a student who has an excess of* $5 - \frac{10}{3} = \frac{5}{3}$ (the fact that this is $\frac{5}{3}$ is an accident, do not let that confuse you). The degree of the cluster is 4. Hence we can view the 1-cluster as a student of value $\frac{5}{3}$ and degree 4.

What to make of the remaining student? We have already used up all of the muffins, so that student can be viewed as needing $\frac{5}{3}$ but not having any muffins. So its value is $-\frac{5}{3}$. Rather than think of negative numbers we instead think of this student as being *a muffin who needs* $\frac{5}{3}$. Note also that this vertex (which now represents a muffin) has degree 4. Hence we can view this as the following Scott-Muffin problem:

$$\text{Scott}\left[\left(1, \frac{5}{3}, 4\right), \left(1, \frac{5}{3}, 4\right)\right].$$

We call it SC(5, 3)-1. It is partially represented by Graph 13.2. The problem is now to put nonnegative weights on the edges such that:

- The weighted degree of the muffin vertex is $\frac{5}{3}$.
- The weighted degree of each student vertex is $\frac{5}{3}$.

We now state a conjecture with two parts. One part we use now, one we use later.

Conjecture 13.4. *If either* (1) *there are no minor students, or* (2) *no set of clusters contains all of the minor students, then all the pieces being given to the major muffins will be the same size.*

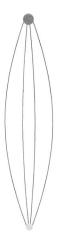

Graph 13.2. Five Muffins, Three Students, SC(5, 3)-1.

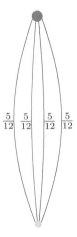

Graph 13.3. Five Muffins, Three Students, Solution to SC(5, 3)-1.

Using this conjecture, and the fact that there are no minor students, the problem is now easy: Assign each edge $\frac{5}{12}$ as in Graph 13.3.

How do we go from the solution of SC(5, 3)-1 to a solution of SC(5, 3)-0? The bottom node is really a cluster of two student-vertices and an internal muffin-vertex. Recall that the weights were how much these students were going to give away. Consider one of these students, Alice. Alice is connected

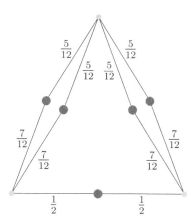

Graph 13.4. Five Muffins, Three Students, Solution to SC(5, 3)-0.

to 2 external muffin-nodes. Using the solution to SC(5, 3)-1 (Graph 13.3) we see that, for each of these muffins, she gives away $\frac{5}{12}$ and hence keeps $\frac{7}{12}$. Hence each student keeps $2 \times \frac{7}{12} = \frac{7}{6}$. They now need to split the internal muffin so that each one gets $\frac{5}{3}$. Hence they each need $\frac{5}{3} - \frac{7}{6} = \frac{1}{2}$. We give the solution to SC(5, 3)-0 in Graph 13.4.

In this case obtaining the solution to the SC(5, 3)-0 from the SC(5, 3)-1 was easy. It is not always so easy and we do not always split internal muffins $(\frac{1}{2}, \frac{1}{2})$.

In summary, we transformed SC(5, 3)-0

$$\text{Scott}\left[(5, 1, 2), \left(1, \frac{5}{3}, 4\right), \left(2, \frac{5}{3}, 3\right)\right]$$

into the easier problem SC(5, 3)-1

$$\text{Scott}\left[\left(1, \frac{5}{3}, 4\right), \left(1, \frac{5}{3}, 4\right)\right].$$

We then solved SC(5, 3)-1 and used its solution to solve SC(5, 3)-0.

13.3. Thirty-Five Muffins, Thirteen Students

We now do the problem of $f(35, 13)$. We will use the ideas from Section 13.2; hence we will assume familiarity with the definitions and ideas presented there. We will need a few new ideas as well.

We assume $f(35, 13) > \frac{1}{3}$. By Theorem 2.6, every muffin is cut into 2 pieces, so there are 70 pieces. We leave it to the reader to show that there are eight 5-students, five 6-students, forty 5-shares, and thirty 6-shares. We express this as the following Scott-Muffin problem:

$$\text{Scott}\left[(35, 1, 2), \left(5, \frac{35}{13}, 6\right), \left(8, \frac{35}{13}, 5\right)\right].$$

We call this problem SC(35, 13)-0. It is somewhat represented by Graph 13.6, though we need to explain why that graph represents the problem, so we will look at Graph 13.5 first.

Since there are thirty 6-shares, the 6-students can only use pieces of 30 muffins. Hence there are 5 muffins that are used entirely by the 5-students.

We will assume that the 5 muffins that are shared by the 5-students form one 1-cluster and two 2-clusters. Graph 13.5 shows those clusters (without the external muffins—that would make a mess) along with the 6-students. It turns out that essentially every muffin problem begins this way: (1) find V so that everyone is either a V-student or a $(V-1)$-student ($V = \lceil \frac{2m}{s} \rceil$), (2) find that the $(V-1)$-students must share m' muffins between them, (3) find an L such that the muffin-sharing can be represented by clusters of length L and $L-1$.

Given the Scott-Muffin problem and the clusters, Graph 13.6 represents it (other graphs might also).

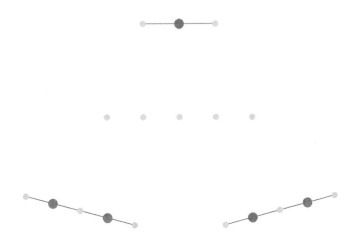

Graph 13.5. Thirty-Five Muffins, Thirteen Students: Clusters.

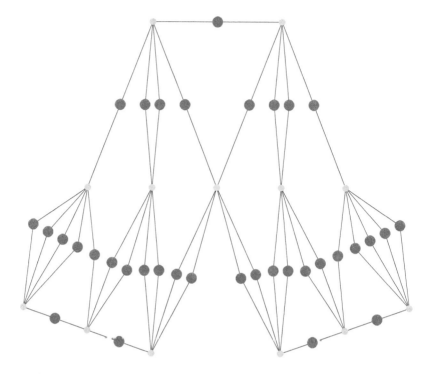

Graph 13.6. Thirty Five Muffins and Thirteen Students, SC(35, 13)-0.

As in Section 13.2, we will transform SC(35, 13)-0 into a smaller problem. Look at Graph 13.6.

(1) The five 6-students will be viewed as not having any muffins adjacent to them (these are the external muffins of the clusters in the next two items) hence these five 6-students need $\frac{35}{13}$ each and have nothing to begin with. These are now viewed as muffins and denoted $(5, \frac{35}{13}, 6)$.

(2) There are two 2-clusters of 5-students (they are at the bottom of both Graphs 13.5 and 13.6). We focus on one of them; however, the same goes for the other one. The three students need $3 \times \frac{35}{13} = \frac{105}{13}$ muffins. The cluster has 2 internal muffins and 11 external muffins for a total of 13 muffins. Hence the cluster becomes a student of value $13 - \frac{105}{13} = \frac{64}{13}$. Note that there are 11 edges coming out of the cluster. Since there are 2 of these clusters we denote this $(2, \frac{64}{13}, 11)$. Note that

$\frac{\text{degree}}{\text{value}} = \frac{11}{64/13} \sim 2.23$. This ratio of degree to value is larger than the one in the next item, so these are the major students.

(3) There is one 1-cluster of 5-students (it is at the top of both Graphs 13.5 and 13.6). The two students need $2 \times \frac{35}{13} = \frac{70}{13}$ muffins. The cluster has 1 internal muffin and 8 external muffins for a total of 9 muffins. Hence the cluster becomes a student of value $9 - \frac{70}{13} = \frac{47}{13}$. Note that there are 8 edges coming out of the cluster. Since there is only 1 of these clusters we denote this $(1, \frac{47}{13}, 8)$. Note that $\frac{\text{degree}}{\text{value}} = \frac{8}{47/13} \sim 2.21$. This ratio of degree to value is smaller than the one in the prior item, so these are the minor students.

Hence we have the following Scott-Muffin problem:

$$\text{Scott}\left[\left(5, \frac{35}{13}, 6\right), \left(2, \frac{64}{13}, 11\right), \left(1, \frac{47}{13}, 8\right)\right].$$

We call this problem SC(35, 13)-1. It is partially captured by Graph 13.7.

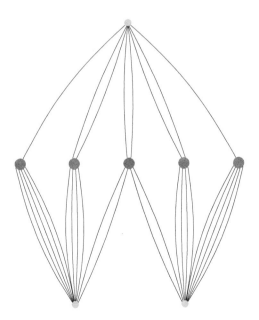

Graph 13.7. Thirty-Five Muffins, Thirteen Students, SC(35, 13)-1.

While we have spilled a lot of ink, all we've done so far is transformed SC(35, 13)-0:

$$\text{Scott}\left[(35, 1, 2), \left(5, \frac{35}{13}, 6\right), \left(8, \frac{35}{13}, 5\right)\right]$$

into SC(35, 13)-1:

$$\text{Scott}\left[\left(5, \frac{35}{13}, 6\right), \left(2, \frac{64}{13}, 11\right), \left(1, \frac{47}{13}, 8\right)\right].$$

We will transform SC(35, 13)-1 to a new problem SC(35, 13)-2. We will then find a solution to SC(35, 13)-2 and use it to find a solution to SC(35, 13)-1, and use that to find a solution to SC(35, 13)-0.

13.3.1. *Transforming* SC(35, 13)-1 *to* SC(35, 13)-2

Recall SC(35, 13)-1:

$$\text{Scott}\left[\left(5, \frac{35}{13}, 6\right), \left(2, \frac{64}{13}, 11\right), \left(1, \frac{47}{13}, 8\right)\right]$$

which is represented by Graph 13.7. Since there are no clusters, by Conjecture 13.4 we give all the major students equal amounts on all of their edges.

The two major students (the two student-vertices at the bottom of Graph 13.7) each want weighted degree $\frac{64}{13}$ and are of unweighted degree 11. Hence we give each of the edges coming out of it weight $\frac{64}{13} \times \frac{1}{11} = \frac{64}{143}$.

With these edges taken care of we will recurse into a smaller Scott-Muffin SC(35, 13)-2. Before defining SC(35, 13)-2 we look at the muffin vertices of SC(35, 13)-1 that have gotten some of the way towards their weighted degree.

There are two kinds of muffin vertices in Graph 13.7. Note that the muffin vertices are the ones in the middle layer. SC(35, 13)-1:

- The 3 muffin-vertices will have 2 edges to the 1 minor student. Since these muffins originally needed weighted degree $\frac{35}{13}$ and now have, from the edges to the major students, $4 \times \frac{64}{143} = \frac{256}{143}$, they now need just $\frac{35}{13} - \frac{256}{143} = \frac{129}{143}$.

- The 2 muffin-vertices will have 1 edge to the 1 minor student. Since these muffins originally needed weighted degree $\frac{35}{13}$ and now have, from the edges to the major students, $5 \times \frac{64}{143} = \frac{320}{143}$, they now need just $\frac{35}{13} - \frac{320}{143} = \frac{5}{11}$.

We will now define the SC(35, 13)-2 problem.

(1) There is 1 muffin of value $\frac{47}{13}$ and degree 8. We denote this as $(1, \frac{47}{13}, 8)$. (This used to be the 1 minor student, which is the top most student in Graph 13.7.)

(2) There are 3 students of value $\frac{129}{143}$ and degree 2. We denote this as $(3, \frac{129}{143}, 2)$. Note that $\frac{\text{degree}}{\text{value}} = \frac{2}{129/143} \sim 2.22$. These students have a larger $\frac{\text{degree}}{\text{value}}$ than those in the next item so these are the major students. (These used to be the muffins that had 2 edges to the minor student.)

(3) There are 2 students of value $\frac{5}{11}$ and degree 1. We denote this as $(2, \frac{5}{11}, 1)$. Note that $\frac{\text{degree}}{\text{value}} = \frac{1}{5/11} \sim 2.20$. These students have a smaller $\frac{\text{degree}}{\text{value}}$ than those in the prior item so these are the minor students. (These used to be the muffins that had 1 edge to the minor student.)

Hence we have

$$\text{Scott}\left[\left(1, \frac{47}{13}, 8\right), \left(3, \frac{129}{143}, 2\right), \left(2, \frac{5}{11}, 1\right)\right].$$

We call this problem SC(35, 13)-2. It is partially captured by Graph 13.8. This graph has no clusters.

We use Conjecture 13.4 and give all the major students equal amounts on all of their edges. Since each major student has degree 2 and must get $\frac{129}{143}$ all of those edges get weight $\frac{129}{143} \times \frac{1}{2} = \frac{129}{286}$. The minor students need $\frac{5}{11}$ and are of degree 1 so the edge to each one must be $\frac{5}{11}$. This completes the solution, though we need to check that the muffins worked out (the first triple in SC(35, 13)-2). For the solution see Graph 13.9.

We need to check that the first part of SC(35, 13)-2 works: $(1, \frac{47}{13}, 8)$. There are 2 edges going into the muffin node of weight $\frac{5}{11}$ and 6 of weight $\frac{129}{286}$.

Graph 13.8. Thirty Five Muffins, Thirteen Students, SC(35, 13)-2.

Hence the total weight going into the muffin is

$$2 \times \frac{5}{11} + 6 \times \frac{129}{286} = \frac{47}{13}.$$

Do not be surprised at this. Because of the way we set it up, it had to be this way.

13.3.2. *Using the Solution to* SC(35, 13)-2 *to Solve* SC(35, 13)-1

We use the solution to SC(35, 13)-2 as expressed in Graph 13.9 to solve SC(35, 13)-1. Actually, this is quite easy, since when we went from SC(35, 13)-1 to SC(35, 13)-2, we had already assigned weights to edges and then removed them. Now all we need to do is put them back. Graph 13.10 shows a solution to SC(35, 13)-1.

13.3.3. *Using the Solution of* SC(35, 13)-1 *to Solve* SC(35, 13)-0

We explain how to take a solution to SC(35, 13)-1 and use it to obtain a solution to SC(35, 13)-0. This will be a case where clusters become vertices and some thought is needed to convert the solution. We will be asking

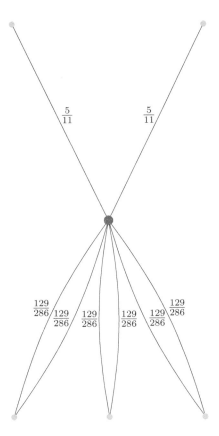

Graph 13.9. Thirty Five Muffins, Thirteen Students, Solution to SC(35, 13)-2.

you to flip back and forth between (1) the problem SC(35, 13)-0 which is Graph 13.6, (2) the solution to SC(35, 13)-1, which is Graph 13.10, and (3) the solution to SC(35, 13)-0, which is Graph 13.11.

The left bottom student-vertex in Graph 13.10 corresponds to the left bottom cluster of Graph 13.6. The 11 edges coming out of the left bottom student-vertex in Graph 13.10 correspond to the 11 2-paths (student-muffin-student) coming out of the left bottom cluster of Graph 13.6. We think of the students in the left bottom cluster as *giving away* $\frac{64}{143}$ and *keeping* $1 - \frac{64}{143}$. Hence we get part of Graph 13.11.

Look at the left most student-vertex in the cluster (the same will hold for the right most) who we call Alice. Alice keeps for herself $4 \times \frac{79}{143} = \frac{316}{143}$ and

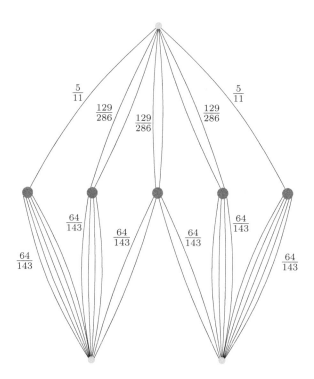

Graph 13.10. Thirty Five Muffins, Thirteen Students, Solution to SC(35, 13)-1.

needs (which will come from the internal muffin) $\frac{35}{13} - \frac{316}{143} = \frac{69}{143}$. Hence the muffin between left and middle is split $\frac{69}{143}$ for left and $1 - \frac{69}{143} = \frac{74}{143}$ for middle.

The same happens for the right most student in the cluster. So now the left and right both have weighted degree $\frac{35}{13}$. What about the middle? He has

$$3 \times \frac{79}{143} + 2 \times \frac{74}{143} = \frac{35}{13}.$$

This should not surprise you. We set it up this way.

The rest of the edges are mostly forced. Look at the 5 student-vertices on the third level from the bottom in Graph 13.11 (don't look at the edge from those nodes going up). Look at the left most student vertex. We already know that the 5 edges coming into it from the bottom contribute $5 \times \frac{64}{143} = \frac{320}{143}$.

Scott Huddleston's Method

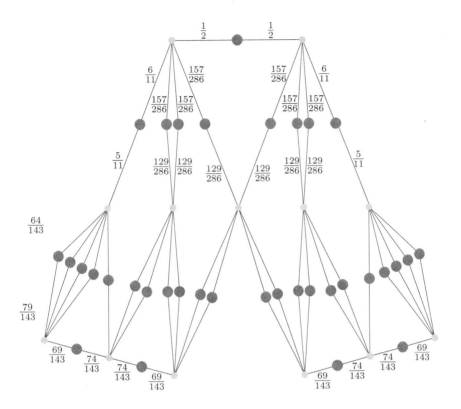

Graph 13.11. Thirty Five Muffins, Thirteen Students, Solution to SC(35, 13)-0.

Since the node needs weighted degree $\frac{35}{13}$, the edge coming out of it going upwards must have weight

$$\frac{35}{13} - \frac{320}{143} = \frac{5}{11}.$$

This edge of weight $\frac{5}{11}$ goes into a muffin-vertex. Since this muffin vertex has weighted degree the other edge coming out of it has weight

$$1 - \frac{5}{11} = \frac{6}{11}.$$

In a similar manner, we can find the weights of all of the edges. It will all work out. See Graph 13.11 for the full solution.

13.4. Reflections on What We Have Done

We have demonstrated a way to, given Scott-Muffin A, find a smaller Scott-Muffin problem B, such that a solution to B gives a solution to A. We do not know that an optimal solution to B gives an optimal solution to A but we believe this to be true. This has held for every single case we have tried. It will be one of our conjectures.

For the problem of finding a (35, 13)-procedure we did the following: Recast it as finding a solution to SC(35, 13)-0:

$$\text{Scott}\left[(35, 1, 2), \left(5, \frac{35}{13}, 6\right), \left(8, \frac{35}{13}, 5\right)\right].$$

Reduced SC(35, 13)-0 to SC(35, 13)-1:

$$\text{Scott}\left[\left(5, \frac{35}{13}, 6\right), \left(2, \frac{64}{13}, 11\right), \left(1, \frac{47}{13}, 8\right)\right].$$

Reduced SC(35, 13)-1 to SC(35, 13)-2:

$$\text{Scott}\left[\left(1, \frac{47}{13}, 8\right), \left(3, \frac{129}{143}, 2\right), \left(2, \frac{5}{11}, 1\right)\right].$$

This last problem, SC(35, 13)-2, was easy (if it wasn't we would have found a SC(35, 13)-3). We got the optimal solution to SC(35, 13)-2 and used it to get a solution to SC(35, 13)-1. We took this solution to SC(35, 13)-1 and used it to get a solution to SC(35, 13)-0, our original problem. By conjecture we have the optimal solution for SC(35, 13)-0.

Carrying out the reduction of a Scott-Muffin problem to a smaller Scott-Muffin problems is very fast. Using a solution for Scott-Muffin problem B to get a solution for Scott-Muffin problem A is also very fast. Note that the reduction in size is often large as well so there are not that many iterations.

13.5. Scott's Algorithm

From Scott's algorithm, we were able to extract six conjectures about the muffin problem with the property that if all six are true, then the algorithm finds optimal solutions.

Essentially, the conjectures characterize what optimal solutions look like. Once you know the conjectures, the algorithm can be understood as a way to find a solution that follows the conjectures. We first present the conjectures, and then present the algorithm. We believe that the conjectures are true but hard to prove.

We will assume that the reader has read the past section and hence knows the following: Scott-Muffin problem, Major Student, Minor Student.

The conjectures and algorithm in this section are about the case where $f(m, s) > \frac{1}{3}$. We will discuss how to (slightly) modify the algorithm to account for what happens if $f(m, s) = \frac{1}{3}$ in Section 13.6. The modification will rely on (what else?) another conjecture.

13.5.1. *The Conjectures*

The first conjecture we have already seen: the V-conjecture. We restate it in the form we will use it.

Conjecture 13.5. *Let $m \geq s$. Let $V = \lceil \frac{2m}{s} \rceil$. There is an optimal (m, s)-procedure where everyone is either a V-student or a $(V - 1)$-student.*

Conjecture 13.6. *In an optimal solution to a Scott-Muffin problem, there are two kinds of muffins.*

- *Minor muffins, which give one piece to one minor student, a second piece to a second minor student, and the rest of their pieces to major students.*
- *Major muffins, which give exactly one piece to one minor student (if there are any minor muffins), and the rest of its pieces to major students.*

Exception: if there are no minor muffins, then major muffins may give all pieces to major students.

Note that the number of major and minor muffins will be determined by the 9 numbers in the Scott-Muffin problem.

Note also that the above conjecture allow us to draw a graph, which we call the *minor muffin graph* (it will be a subgraph of the graphs we had in the last section). In this graph, the nodes are minor students, and two minor students have an edge between them if they both receive a piece from the same minor muffin. The next conjecture characterizes what this graph looks like.

Conjecture 13.7. *Assume we are looking at a Scott-Muffin problem. In an optimal solution, the minor muffin graph consists of clusters of length L and L − 1 for some number L. A cluster in a minor muffin graph is defined below.*

Definition 13.8. A *cluster of a minor graph* is a set of l nodes connected in a line by $l - 1$ edges. Note that this definition corresponds to the definition of cluster given in the earlier examples. The edges of a cluster in a minor graph correspond to the internal muffins of the muffin–student–muffin–$\cdots$–student–muffin sequence, since those internal muffins are minor muffins. The other muffins attached to the students in the sequence are all major muffins, so they are not drawn in the minor muffin graph.

Lemma 13.9. *Assume Conjecture 13.7 is true. Assume we are looking at a Scott-Muffin problem with parameters as in Definition 13.1.*

(1)
$$L = \left\lceil \frac{1}{\frac{n_m}{n_{s2}} + 1 - p_{s2}} \right\rceil. \qquad (13.3)$$

(2) *There are*
$$a = L n_m - L p_{s2} n_{s2} + L n_{s2} - n_{s2} \qquad (13.4)$$

$L - 1$ *clusters.*

(3) *There are*
$$b = \frac{n_{s2} - a(L - 1)}{L} \qquad (13.5)$$

L-*clusters.*

(4) *The exception from Conjecture 13.6 applies exactly when $L = 1$.*

(5) *It is possible to partition the graph into L and $L - 1$ clusters exactly when there are fewer edges than vertices in the minor graph, that is, when $n_{s2} p_{s2} < n_m + n_{s2}$.*

Proof. We just note that the calculation of a and b uses the following:

Each $L-1$-cluster uses $L-1$ minor students and $p_{s_2}(L-1)-L+2$ muffins, and each L-cluster uses L minors and $p_{s_2}L-L+1$ muffins. Using this information, we can calculate exactly how many L and $L-1$ clusters there are.

For (5), it is clear that if there are more edges than vertices, the minor graph must have a cycle and therefore is not a forest of clusters. On the other hand, if there are N vertices and E edges, then there is an $L \geq 1$ such that $N\frac{L-1}{L} \geq E \geq N\frac{L-2}{L-1}$. So, we can construct a minor graph with N vertices and E edges which consists solely of L-clusters and $(L-1)$-clusters. Finally, note that the minor graph has n_{s_2} vertices and $n_{s_2}p_{s_2} - n_m$ edges. □

Conjecture 13.10. *Assume that a solution to a Scott-Muffin problem conforms to Conjectures 6.12, 13.6 and 13.7. Then, for each muffin, the pieces that it gives to minors are bigger than the pieces it gives to majors.*

The consequence of this is that we don't have to think about which minor students receive pieces—only which cluster of them receives pieces.

Conjecture 13.11. *If there are too many muffins to be able to form clusters, then the optimal solution will involve giving each major student pieces which are all the same size. Furthermore, that size will be the smallest piece necessary in the procedure. (That second clause is not necessary to know that the algorithm yields optimal solutions, but is true in all cases checked so far.)*

By the lemma, the conjecture kicks in when $n_{s_2}p_{s_2} \geq n_m + n_{s_2}$.

Conjecture 13.12. *If the condition of the last conjecture holds, then the muffins should give pieces to the majors as evenly as possible.*

This means that the muffins will each give either P or $P-1$ pieces to the major students, where

$$P = \left\lceil \frac{n_{s_1}p_{s_1}}{n_m} \right\rceil. \tag{13.6}$$

There will be $x = n_{s_1} p_{s_1} - (P-1)n_m$ muffins which give P pieces, and $y = n_m - a$ muffins which give $P-1$ pieces.

Theorem 13.13. *If the above conjectures are true, then the algorithm described in the next section always finds optimal solutions.*

We will see that this is true.

13.5.2. The Algorithm

The algorithm will input a Scott-Muffin problem and output a α such that $f(m,s) \geq \alpha$ and the procedure that proves $f(m,s) \geq \alpha$. There are three cases depending on the input. In each case, we use the conjectures to determine how the algorithm must proceed.

13.5.2.1. Case 1: No majors or No minors

Without loss of generality suppose that there are no minors, so $n_{s_2} = 0$.

We must have then that the total value of muffins and majors is the same,

$$n_m v_m = n_{s_1} v_{s_1} \tag{13.7}$$

and that the total pieces are the same,

$$n_m p_m = n_{s_1} p_{s_1}. \tag{13.8}$$

Dividing those two equations, we get

$$\frac{v_m}{p_m} = \frac{v_{s_1}}{p_{s_1}}. \tag{13.9}$$

If we split all muffins into even sized pieces, then every piece is of size $\frac{v_m}{p_m}$. Each student will get p_{s_1} of these pieces. Because of the last equation, this is a valid procedure. It is also clearly an optimal procedure.

13.5.2.2. Case 2: We can make clusters, or $n_{s_2} p_{s_2} < n_m + n_{s_2}$

In this case, according to the conjecture, an optimal solution must have major and minor muffins divided into L- and $(L-1)$- clusters. Also according to conjectures, there are a $(L-1)$-clusters and b L-clusters.

Let's recap the situation. We have n_{s_1} majors, who receive all of their pieces from major muffins. We also have n_{s_2} minors who receive pieces from SPECIFIC major and minor muffins, depending on their place in their cluster. Which muffins they receive pieces from is specified by the way that we drew the minor graph. Therefore, **the only question that remains is which major muffins give pieces to which majors**.

However, all major muffins are linked to either an L-cluster or an $(L-1)$-cluster. Furthermore, we can throw out some information. Suppose that in an optimal solution, major s_1 gets a piece from major muffin m_1, and major s_2 gets a piece from major muffin m_2. Further, suppose that m_1 and m_2 are connected to the same cluster. Then, we could instead let s_1 get their piece from m_2 and s_2 get their piece from m_1, and by Conjecture 13.4 the solution would still remain optimal.

Therefore, the only piece of information that we need to find is: **for each major student, which clusters are the major muffins that it gets its pieces from connected to?**

In order to figure this out, we make use of a very clever discovery that Scott made. The question of which major muffins get their pieces from which clusters is exactly like a Scott-Muffin problem! We will construct a *new muffin problem*, whose solution will tell us the solution to the *old muffin problem*.

In order to construct this new muffin problem, for each old major make a new muffin. For each old L or $(L-1)$-cluster, make a new student. Finally, let s and c be an old major and old cluster, respectively. Let m' and s' be the corresponding new muffin and new student, respectively. Then, for every piece that s gets from a major muffin connected to c, let m' give a piece of the same size to s'.

This transformation to a new muffin problem has the property that if we find an optimal procedure for the new muffin problem, and then use that procedure to decide where old majors get their pieces in the old problem, then that will yield an optimal solution in the old problem as well.

In order to find out which major muffins give pieces to which clusters, we run Scott's algorithm again on the Scott-Muffin problem below. You can use algebra to find the V and P for each cluster. The result is the following mess:

$$(n_{s_1}, v_{s_1}, p_{s_1}),$$
$$(a, (Lp_{s_2} - (L-1))v_m - Lv_{s_2}, p'_{s_1}), \tag{13.10}$$
$$(b, ((L-1)p_{s_2} - (L-2))v_m - (L-1)v_{s_2}, p'_{s_2}),$$

where

$$\begin{aligned} p'_{s_1} &= (Lp_{s_2} - 2(L-1))(p_m - 1) + (L-1)(p_m - 2), \\ p'_{s_2} &= ((L-1)p_{s_2} - 2(L-2))(p_m - 1) + (L-2)(p_m - 2). \end{aligned} \tag{13.11}$$

13.5.2.3. Case 3: We can't make clusters, or $n_{s_2} p_{s_2} \geq n_m + n_{s_2}$

In this case, according to the second part of Conjecture 13.5, we must divide the majors evenly. We have already found $f(m, s)$ and if all we want is the answer we could stop. However, in order to find the entire procedure, we need to continue.

In addition, according to Conjecture 13.6, we know that there are x muffins which give P pieces to majors, and y muffins which give $P - 1$ pieces to majors.

Therefore, the only question that remains is which minors get pieces from which muffins. There are two kinds of muffins, and the only students we don't know about are minors, so we will make a Scott-Muffin problem where the old minors become the new muffins, and the old muffins become the new majors and minors.

Therefore, we recurse Scott's algorithm with the following problem:

$$(n_{s_2}, v_{s_2}, p_{s_2}),$$
$$\left(x, v_m - \frac{Pv_{s_1}}{p_{s_1}}, p_m - P\right), \tag{13.12}$$
$$\left(y, v_m - \frac{(P-1)v_{s_1}}{p_{s_1}}, p_m - (P-1)\right).$$

The parameters are decreasing and will eventually get to a point where there are no minor muffins.

13.6. What if $f(m, s) = \frac{1}{3}$?

The following is a fact, not a conjecture:

Fact: Scott's algorithm will find a procedure where every muffin is cut into exactly two pieces.

Should Scott's algorithm output a procedure with smallest piece $< \frac{1}{3}$ then there are two ways to modify the algorithm:

- Run FINDPROC($m, s, \frac{1}{3}$) to find a procedure with smallest piece $\frac{1}{3}$.
- Use Theorem 4.5 to obtain a procedure with smallest piece $\frac{1}{3}$.

It is known that if $m \geq s$ then $f(m, s) \geq \frac{1}{3}$. There is a proof of this on the MUFFIN website and also a proof in Richard's paper [Chatwin (2019)].

Appendix A

Math Notation

In this chapter, we present some math notation that is used throughout the book.

A.1. Sets

A *set* is a collection of objects. For us it will always be a collection of numbers. No element can appear twice in a set.

Notation A.1. Let A be a set and x be a number.

(1) $x \in A$ means x is in the set A.
(2) $x \notin A$ means x is not in the set A.

We denote sets in several ways:

(1) If the set is finite and small we can just list out the elements. For example,
$$\left\{\frac{5}{12}, \frac{6}{12}, \frac{7}{12}\right\}.$$

(2) If the set is finite and large but has a pattern, we can list out the elements so that the pattern is clear, and use "...". For example,
$$\left\{\frac{13}{256}, \frac{14}{256}, \frac{15}{256}, \ldots, \frac{243}{256}\right\}.$$

(3) If the set is finite and large but has no obvious pattern, then we can describe the set in English (and later with math notation). For example,
$$\{x \in \mathbb{N} : x \leq 100 \text{ and } x \text{ can be written as the sum of 3 squares}\}.$$

(4) If the set is infinite but has a pattern, we can list out the elements so that the pattern is clear, and use "...". For example,

$$\{3, 6, 9, \ldots\},$$

$$\{\ldots, -3, -2, -1, 0, 1, 2, 3, \ldots\}.$$

(5) If the set is infinite but has no obvious pattern, then we can describe the set in English (and later with math notation). For example,

$$\{x \in \mathbb{N} : x \text{ can be written as the sum of 3 squares}\}.$$

(6) Some sets have well-established names:
   - $\emptyset$ is the set with no elements. It is also denoted { }.
   - $\mathbb{N}$ is the set of naturals, which is $\{0, 1, 2, \ldots\}$. Some textbooks do not include 0. They are wrong.
   - $\mathbb{Z}$ is the set of integers, which is $\{\ldots, -2, -1, 0, 1, 2, \ldots\}$.
   - $\mathbb{Q}$ is the set of rationals, which is $\{\frac{a}{b} : a, b \in \mathbb{Z}, b \neq 0\}$.
   - $\mathbb{R}$ is the set of reals. The reals are actually hard to define formally so we take the intuitive definition of finite and infinite decimal expansions. There are reals that are not rationals. It is known that $\sqrt{2}$ is a real but not a rational.
   - (a, b) is the set $\{x \in \mathbb{R} : a < x < b\}$. Note that a, b are not included. This is called *an open interval*.
   - $[a, b]$ is the set $\{x \in \mathbb{R} : a \leq x \leq b\}$. Note that a, b are included. This is called *a closed interval*.
   - $[a, b)$ and $(a, b]$ we leave to you to define. These are called *clopen intervals*.

Notation A.2. Let A and B be sets.

(1) $A \cup B$, pronounced A *union* B, is the set of elements in A *or* B. For example,

$$\{1, 3, 4, 5\} \cup \{1, 5, 10, 15\} = \{1, 3, 4, 5, 10, 15\}.$$

(2) $A \cap B$, pronounced A *intersect* B, is the set of elements in A *and* B. For example,

$$\{1, 3, 4, 5\} \cap \{1, 5, 10, 15\} = \{1, 5\}.$$

(3) A set with W elements in it is called *a W-set*.
(4) $A \subseteq B$, pronounced *A is a subset of B*, means that every element of A is an element of B.
(5) $A \supseteq B$, pronounced *A is a superset of B*, means that every element of B is an element of A.
(6) A subset of A with W elements is called *a W-subset of A*. If A is understood then we just use the term *W-subset*.
(7) Assume $A \subseteq \mathbb{Q}$. Moreover, assume (informally) that all we care about is $\mathbb{Q}$. Then the *complement of A is the set of elements of $\mathbb{Q}$ that are not in A*. We denote this by $\overline{A}$. More generally, $\overline{A}$ is defined when we have a universe of discourse (in the case above, $\mathbb{Q}$) and we take the complement relative to it.
(8) $A - B$ is the set of elements in A that are not in B. Formally this is $A \cap \overline{B}$.
(9) $A \times B$ is the set of all ordered pairs (a, b) such that $a \in A$ and $b \in B$. We leave the definition of $A \times B \times C$, and beyond, to the reader.
(10) $A^2 = A \times A$. $A^n = A \times \cdots \times A$ (A appears n times).

Exercise A.3. Let

$$A = \{2, 4, 6, \ldots, 100\},$$
$$B = \{3, 6, 9, \ldots, 99\}.$$

(1) List all of the elements of $A \cap B$.
(2) Write $A \cap B$ in the form $\{x \in \mathbb{N} : \text{BLAH}\}$.
(3) List all of the elements of $A \cup \emptyset$.
(4) List all of the elements of $A \cap \emptyset$.

Definition A.4.

(1) A *multiset* is a set where we allow elements to appear more than once. For example, $\{1, 1, 2, 3\}$ is a multiset.
(2) Let A be a set. Then B is a *multisubset of A* if every element of B is in A and an element of A can appear many times in B. If it is understood that B is a multiset we just use the term *subset of A*.
(3) Let B be a multiset. B is a *W-multisubset of A* if B is a multisubset of A that has W elements in it. Note that the set $\{1, 1, 3, 4\}$ has just 4

elements in it. If it is understood that B is a multiset we use the term W-*subset*. We will use this in Chapter 5 with $W = V$ and $W = V - 1$.

Example A.5. Let $B = \{5, 6, 7\}$. Then both $\{5, 7\}$ and $\{6, 6\}$ are 2-subsets of B. Also, $\{6, 6, 7\}$ is a 3-subset of B.

Example A.6. Let

$$A = \{1, 2, 3, 4, 10\}.$$

Then the following are subsets of A. The first one is a multiset but we still call it a *subset of A* since calling it a *multisubset* is clumsy.

(1) $\{1, 1, 1, 2, 3\}$.
(2) $\{1, 2, 10\}$.

A.2. Functions and Bijections

Let A and B be sets. A *function f from A to B* is a rule that takes an element of A and maps it to an element of B. For example,

$$f(x) = 2x + 1$$

is a function from $\mathbb{N}$ to $\mathbb{N}$. It can also be regarded as a function from $\mathbb{N}$ to the odd numbers.

Notation A.7. Let A and B be sets and let f be a function from A to B.

(1) A is called the *domain* of f, and B is called the *codomain* of f.
(2) f is *injective* (also called one-to-one) if

$$(\forall x, y \in A)\, [x \neq y \to f(x) \neq f(y)].$$

In other words, no two elements of A map to the same element of B.

Example A.8.

(a) If the domain and codomain are both $\mathbb{N}$ then $f(x) = x^2$ is injective: If $x^2 = y^2$ then $(x - y)(x + y) = 0$ so either $x = y$ or $x = -y$. If $x = -y$ then, since $x, y \geq 0$, we must have $x = y = 0$.

(b) If the domain is $\mathbb{Z}$ and the codomain is $\mathbb{N}$ then $f(x) = x^2$ is not injective since, for example, $f(3) = f(-3) = 9$.

(3) f is *surjective* (also called onto) if

$$(\forall y \in B)\, (\exists x \in A)\, [f(x) = y].$$

In other words, every element in the codomain gets mapped to by an element of the domain.

Example A.9.

(a) If the domain is $\mathbb{R}$ and the codomain is $\mathbb{Z}$, then $f(x) = \lceil x \rceil$ is surjective: if $y \in \mathbb{Z}$, then take $x = y$ to get $f(x) = y$.
(b) If the domain is $\mathbb{Z}$ and the codomain is $\mathbb{Q}$, then $f(x) = x + 1$ is not surjective. No element of $\mathbb{Z}$ maps to $\frac{1}{2}$.

(4) f is *bijective* (also called one-to-one and onto) if it is both injective and surjective.

Example A.10.

(a) If the domain and the codomain are both $\mathbb{N}$, then $f(x) = x^2$ is not bijective since it is not surjective. No element of $\mathbb{N}$ maps to 2.
(b) If the domain is $\mathbb{R}$ and the codomain is $\mathbb{Z}$, then $f(x) = \lceil x \rceil$ is not bijective since it is not injective. Many elements map to 1. In fact, every element in $(0, 1]$ maps to 1.
(c) If the domain and range are both $(0, 1)$, then $f(x) = 1 - x$ is a bijection. Note that this is the buddy function from Definition 1.12.

Notation A.11. If f is a function from A to B and $C \subseteq A$, then

$$f(C) = \{f(x) : x \in C\}.$$

We now state an important property of bijections. We first need some notation.

Notation A.12.

(1) If A is a finite set, then $|A|$ is the number of elements in A.

(2) if A is a set of intervals, then $|A|$ is the number of shares in A. This notation is not standard. It is likely that it is only used in the mathematics of muffins.

Theorem A.13. *Let f be a bijection A to B.*

(1) *If A is finite, then $|f(A)| = |A|$ (using Notation A.12.1).*
(2) *If A is a set of intervals, then $|f(A)| = |A|$ (using Notation A.12.2).*

If f is a function from A to B, $C \subseteq A$, and C is finite, then $|f(C)| = |C|$.

A.3. Mod Arithmetic

Notation A.14. Let $a, b, s \in \mathbb{N}$. Then

$$a \equiv b \quad (\text{mod } s)$$

means that $a - b$ is divisible by s. This notation is usually only used when $0 \le b \le s - 1$. In that case, b is the remainder when a is divided by s.

Example A.15. $100 \equiv 2 \pmod 7$ since 7 divides $100 - 2 = 98$. Alternatively, 2 is the remainder when 100 is divided by 7.

Example A.16. Let's look at mod 3:
$0 \equiv 0 \pmod 3$.
$1 \equiv 1 \pmod 3$.
$2 \equiv 2 \pmod 3$.
$3 \equiv 0 \pmod 3$.
$4 \equiv 1 \pmod 3$.
$5 \equiv 2 \pmod 3$.

Definition A.17. $a \pmod b$ means the remainder when a is divided by b. For example,

$$29 \bmod 4 = 1.$$

Definition A.18. Let n be $d_L d_{L-1} \cdots d_0$ in base 10.

(1) $\text{sum}(n) = d_0 + d_1 + \cdots + d_L$.

Exercise A.19.

(1) For $700 \leq n \leq 710$ compute
- $n \pmod 3$
- $\text{sum}(n) \pmod 3$

(You should get that $n \equiv \text{sum}(n) \pmod 3$.)

(2) For $0 \leq i \leq 10$ compute $10^i \pmod 3$. (*Advice*: once you know that $10^i \equiv a \pmod 3$, compute $10^{i+1} = 10 \times 10^i \equiv 10a \pmod 3$.)

(3) Spot a pattern in the $10^i \pmod 3$ sequence and make a formula for $10^i \pmod 3$.

(4) Prove that, for all n, $n \equiv \text{sum}(n) \pmod 3$. (*Hint*: If $n = d_L \cdots d_0$ then $n = 10^L d_L + \cdots + 10^1 d_1 + d_0$.)

(5) For $0 \leq i \leq 10$ compute $10^i \pmod 9$. (*Advice*: once you know that $10^i \equiv a \pmod 9$, compute $10^{i+1} = 10 \times 10^i \equiv 10a \pmod 9$.)

(6) Spot a pattern in the $10^i \pmod 9$ sequence and make a formula for $10^i \pmod 9$.

(7) Prove that, for all n, $n \equiv \text{sum}(n) \pmod 9$.

Solution to Exercise A.19

(1) $700 = 3 \times 233 + 1 \equiv 1 \pmod 3$.
$\text{sum}(700) = 7 \equiv 1 \pmod 3$.

$701 = 3 \times 233 + 2 \equiv 2 \pmod 3$.
$\text{sum}(701) = 7 + 1 = 8 \equiv 2 \pmod 3$.

$702 = 3 \times 233 + 3 \equiv 0 \pmod 3$.
$\text{sum}(702) = 7 + 2 = 9 \equiv 0 \pmod 3$.

We omit the rest.

(2) $10^0 = 1 \equiv 1 \pmod 3$.
$10^1 = 10 \equiv 1 \pmod 3$.
$10^2 = 10 \times 10 \equiv 1 \times 1 \equiv 1 \pmod 3$.
$10^3 = 10^2 \times 10 \equiv 1 \times 1 \equiv 1 \pmod 3$.

We omit the rest.

(3) The pattern is that, for all i, $10^i \equiv 1 \pmod 3$.

(4) All $\equiv$ are mod 3.

$$\begin{aligned}n &= d_L \cdots d_0 \\ &= 10^L d_L + \cdots + 10^1 d_1 + d_0 \\ &\equiv 1 \times d_L + \cdots + 1 \times d_1 + d_0 \\ &= \mathrm{sum}(n).\end{aligned}$$

(5) Omitted, but similar to the mod 3 case.
(6) The pattern is that for all i, $10^i \equiv 1 \pmod 9$.
(7) Omitted, but similar to the mod 3 case.

We have established the following:

(1) For all n, $n \equiv \mathrm{sum}(n) \pmod 3$.
(2) For all n, $n \equiv \mathrm{sum}(n) \pmod 9$.

These facts give a very easy way to tell if a number n is divisible by 3 or 9: just sum the digits of n mod 3 or 9 and see if the answer is 0. More information can be obtained, namely what n is congruent to mod 3 or 9.

The fact that $n \equiv \mathrm{sum}(n) \pmod 9$ was often used to check sums before calculators were in wide spread use. For example, if you did the calculation

$$1233987 + 59201 = 1293188$$

you can check it by seeing if

$$1233987 + 59201 \equiv 1293188 \pmod 9$$

$$\mathrm{sum}(1233987) + \mathrm{sum}(59201) \equiv \mathrm{sum}(1293188) \pmod 9$$

$$33 + 17 \equiv 32 \pmod 9.$$

Rather than do the addition we can again pass to the sums.

$$\mathrm{sum}(33) + \mathrm{sum}(17) \equiv \mathrm{sum}(32) \pmod 9$$

$$6 + 8 \equiv 5 \pmod 9$$

$$14 \equiv 5 \pmod 9,$$

which is true.

In reality, the above is not what you would do. Look at

sum(1233987). You would really mod 9 as you go like this:
1+2+3+3 OH, that is, $9 \equiv 0$. So we have 0.
$0 + 9$ OH, that is, $9 \equiv 0$.
$0 + 0 + 8 + 7 = 15 \equiv 6$.

The above is the basis for the procedure known as *casting out 9's*. Before modern computers were in common use, people would check $x + y = z$ for large x, y, z. One way to check the sum was to (in our terminology) make sure that

$$x \pmod 9 + y \pmod 9 = z \pmod 9$$

This was easy to do by doing a running sum of the digits of x and y and z mod 9. If you got a $\neq$ then the sum is wrong. If you got an $=$ then the sum could still be wrong but the chance of that is much less.

We just described a method to check *sums* using mod 9. The same technique can be used to check *products*.

Let's look back at divisibility. Let $n = n_d \cdots n_0$. The following are true:

(1) $n \equiv \text{sum}(n) \pmod 3$.
(2) $n \equiv \text{sum}(n) \pmod 9$.
(3) $n \equiv n_0 \pmod 2$.
(4) $n \equiv n_0 \pmod 5$.

Exercise A.20. The four facts above give easy tricks for determining if a number is divisible by 3, 9, 2, 5. Find tricks for divisibility by 4, 6, 7, 8, 11. (*Hint*: For mod 4 look at $10^i \pmod 4$ and try to find a pattern. Similar for mod 6, mod 7, mod 8, and mod 11.)

A.4. Quantifiers

The symbol $\exists x$ means *there exists x*. When it is used there is a domain in mind that x comes from. For example, if the domain is $\mathbb{N}$, then

$$A = \{x : (\exists y)(\exists z)[x = y^2 + z^2]\}$$

is the set of all natural numbers that can be written as the sum of two squares.

The symbol $\forall x$ means *for all x*. And again, there is a domain. For example, if the domain is $\mathbb{N}$ then

$$A = \{x : (\forall y)[\text{ if } y \text{ divides } x \text{ then } y = x \text{ or } y = 1]\}$$

is the set of all natural numbers that only have 1 and themselves as divisors. These are also called *the primes*.

The notation

$$(\forall 0 \leq n \leq 19)$$

means for all n that are between 0 and 19 inclusive. For example, if the domain is $\mathbb{N}$, then

$$(\forall 0 \leq n \leq 22)(\exists x_1, \ldots, x_8)[n = x_1^3 + \cdots + x_8^3]$$

means that every natural number between 0 and 22 inclusive can be written as the sum of 8 cubes (this is true).

A.5. Summation Notation

We want to write

$$1 + 3 + 5 + \cdots + 99$$

more compactly. We use the notation

$$\sum_{i=0}^{49} 2i + 1.$$

More generally

$$\sum_{i=0}^{n} a_i = a_0 + a_1 + \cdots + a_n.$$

Let $A = \{1, 4, 5, 10\}$. Then

$$\sum_{i \in A} i = 1 + 4 + 5 + 10.$$

We can also write
$$\sum_{i \in A} 2i + 1 = (2 \times 1 + 1) + (2 \times 4 + 1) + (2 \times 5 + 1) + (2 \times 10 + 1).$$
More generally, if A is a set and f is a function with domain A, and $A = \{a_1, \ldots, a_L\}$ then
$$\sum_{i \in A} f(i) = f(a_1) + \cdots + f(a_L).$$

Appendix B

Fair Division

There is a wonderful field called *fair division* which our work can be considered a close cousin of. We describe it briefly.

There is a vast literature on the following problem (stated informally): How can n students, with different tastes, divide a cake *fairly*, in the best way [Aziz and Mackenzie (2016); Brams and Taylor (1995, 1996); Edmonds and Pruhs (2011); Even and Paz (1984); Robertson and Webb (1998)]. In those problems, there is *one* cake and the students may have *different tastes*. For example, Bill prefers chocolate and everyone else prefers kale (Bill is right, the others are wrong).

One can vary the notions of fairness and also look at discrete objects (e.g., how to split an inheritance) as well as continuous (our cake). One can also vary what kind of procedure are allowed: discrete (e.g., Alice cuts, Bob chooses) or continuous (e.g., moving knife protocols where a knife is going over a cake until someone yells stop).

Convention B.1. Everyone thinks that the whole cake has value 1.

We give some definitions and state some theorems.

Definition B.2. Let $n \in \mathbb{N}$. A division of a cake among n students (who may have different tastes) is *proportional* if each person thinks they have $\geq \frac{1}{n}$.

You might think that a proportional division is fair. But consider the following scenario: Alice, Bob, and Carol have split a cake such that:

- Alice thinks (1) she has $\frac{2}{5}$, (2) Bob has $\frac{1}{10}$, (3) Carol has $\frac{1}{2}$. Alice thinks Carol has more than her!
- Bob thinks (1) he has $\frac{2}{5}$, (2) Alice has $\frac{2}{5}$, (3) Carol has $\frac{1}{5}$. Bob is happy.
- Carol thinks (1) she has $\frac{2}{5}$, (2) Alice has $\frac{2}{5} + \frac{1}{10^{10}}$, (3) Bob has $\frac{1}{5} - \frac{1}{10^{10}}$. Carol thinks that Alice has more than her!

Definition B.3. Let $n \in \mathbb{N}$. A division of a cake among n students (who may have different tastes) is *envy free* if each person thinks they have the most (or tied for the most).

We list some of what is known about discrete protocols for cake cutting.

(1) For $n = 2$ there is a well-known protocol that is envy free: Alice cuts the cake in two, and Bob chooses one of the pieces.
(2) There is a proportional cake cutting protocol for n students that takes roughly $n \log n$ cuts [Even and Paz (1984)].
(3) Any proportional cake-cutting protocol for n students requires roughly $n \log n$ cuts [Edmonds and Pruhs (2011)].
(4) There is an envy-free protocol for 3 students where the number of cuts is 5 [Brams and Taylor (1995)].
(5) For all $n \geq 4$ there is an envy-free protocol for n students where the number of cuts is unbounded [Brams and Taylor (1995)]: for all numbers M, there is a way to set up the tastes of the n students so that the number of cuts is at least M. Is there a bounded protocol? See next point.
(6) There is a procedure with $g(n)$ cuts to divide a cake among n students in an envy-free way where $g(n)$ is a very fast growing function [Aziz and Mackenzie (2016)].

Appendix C

$f(m,s)$ Exists! $f(m,s)$ is Rational! $f(m,s)$ is Computable!*

C.1. Introduction

Does $f(m, s)$ always exist? It is plausible that, for any $(24, 11)$-procedures, there is a better one. For example, we have the following:

(1) There is a $(24, 11)$-procedure with smallest piece $\frac{19}{44} - \frac{1}{10}$.
(2) There is a $(24, 11)$-procedure with smallest piece $\frac{19}{44} - \frac{1}{100}$.
(3) There is a $(24, 11)$-procedure with smallest piece $\frac{19}{44} - \frac{1}{1000}$.
(4) Etc., but also:
(5) For all $(24, 11)$-procedures there is some piece $< \frac{19}{44}$.

If that happened then $f(24, 11)$ would not exist.

Do not worry. We give three proofs that $f(m, s)$ exists.

Is $f(m, s)$ always rational? It is plausible that $f(24, 11) = \frac{\pi}{7}$.

Do not worry. We give three proofs that $f(m, s)$ is rational.

Is there a program that will, given m, s, output $f(m, s)$? It is plausible that $f(m, s)$ is not computable.

Do not worry. We give two proofs that $f(m, s)$ is computable.

C.2. Roadmap

Linear Programming and Mixed Integer Programming (which we will define rigorously) are problems that are *known* to be solvable and always have a rational solution.

Our first proof will phrase the problem of determining $f(m, s)$ as solving an insane number of linear programming problems. We stress that this is a proof that $f(m, s)$ exists, is rational, and is computable, but **not** a feasible way to find $f(m, s)$.

Our second proof will phrase the problem of determining $f(m, s)$ as a mixed integer programming problem. This proof that $f(m, s)$ exists, is rational, and computable, can be turned into an algorithm to find $f(m, s)$ for small values of m, s. After the proof we give advice on how to make this algorithm faster and hence perhaps suitable for larger values of m, s.

Our third proof uses elegant (though advanced) ideas from analysis to show that $f(m, s)$ exists and is rational. It does not give a way to actually compute $f(m, s)$.

C.3. Linear and Integer Programming

We urge the reader to look up the definitions of vectors, matrices, not product, and matrix multiplication before proceeding.

Definition C.1. *Linear Programming (LP)* is the following problem: Given a matrix M and vectors $\vec{b}, \vec{c}$ where all of the entries are rational, find a vector $\vec{x}$ such that

- $M\vec{x} \le \vec{b}$;
- $\vec{x} \cdot \vec{c}$ is maximized.

We can replace *maximized* with *minimized*. It may be the case that no such $\vec{x}$ exists. This may happen if the region

$$\{\vec{x} : M\vec{x} \le \vec{b}\}$$

is empty or unbounded. Also, note that equality constraints can be used in addition to inequality constraints. If we want the constraint $\vec{m} \cdot \vec{x} = b$, we can get that by combining the constraints $\vec{m} \cdot \vec{x} \le b$ and $-\vec{m} \cdot \vec{x} \le -b$.

We state the following well-known result without proof.

Lemma C.2.

(1) *Let $(M, \vec{b}, \vec{c})$ be an LP such that the region $\{x : M\vec{x} \le \vec{b}\}$ is nonempty and bounded. Then there exists a rational solution.*

(2) *The following function is computable*: Given an LP $(M, \vec{b}, \vec{c})$, determine if it has a solution, and if it does then output it.

Note C.3. The *Simplex Method*, invented by George Dantzig in the late 1940s, is a well known and often used algorithm to solve LPs. You can find it in any operations research textbook and also on the web. For most inputs it is very fast; however, there are some inputs where it is known to take a long time—formally time 2^n where n is the number of variables. In 1979, Leonid Khachiyan presented a polynomial time algorithm for LP; however, it was slow in practice. In 1984, Narendra Karmarkar presented a polynomial time algorithm for LP which seems to be fast in practice. There are packages for the LP that solve it quickly most of the time.

Definition C.4. A *Mixed Integer Program* (*MIP*) is the following problem: Given a matrix M and vectors $\vec{b}, \vec{c}$ where all of the entries are rational, and a finite set $I \subseteq \mathbb{N}$, find a vector $\vec{x}$ such that:

- $M\vec{x} \leq \vec{b}$.
- $\vec{x} \cdot \vec{c}$ is maximized.
- For every $i \in I$, $x_i \in \mathbb{Z}$. These x_i are called *the integer variables*.

We can replace *maximized* with *minimized*. It may be the case that no such $\vec{x}$ exists. This may happen if the region

$$\{\vec{x} : M\vec{x} \leq \vec{b}\}$$

is empty or unbounded or has no integers for an integer variable.

We state the following well-known result without proof.

Lemma C.5.

(1) *Let $(M, \vec{b}, \vec{c}, I)$ be an MIP such that the region $\{x : M\vec{x} \leq \vec{b}\}$ is bounded. Then there exists a rational solution.*
(2) *The following function is computable*: Given MIP $(M, \vec{b}, \vec{c}, I)$, determine if it has a solution, and if it does then output it.

Note C.6. There is no known method for solving MIPs that runs quickly. Such an algorithm is unlikely (formally, the problem is NP-complete). There are packages for the problem that do well if the dimensions of the matrix are not too large.

C.4. A Proof That Uses Linear Programming

Theorem C.7.

(1) *For all m, s, $f(m, s)$ exists.*
(2) *For all m, s, $f(m, s) \in \mathbb{Q}$.*
(3) *The function $f(m, s)$ is computable.*

Proof. Consider the following (failed) attempt to phrase the muffin problem as an LP.

MUFFINLP

(1) The variables are x_{ij} where $1 \le i \le m$ and $1 \le j \le s$. The intent is that x_{ij} is the fraction of muffin i that student j gets.
(2) For all $1 \le i \le m, 1 \le j \le s, 0 \le x_{ij} \le 1$.
(3) Let $1 \le i \le m$. Consider the ith muffin. The sum of what it gives to student 1, student 2, ..., student s is 1. Hence we have requirement $\sum_{j=1}^{s} x_{ij} = 1$.
(4) Let $1 \le j \le s$. Consider the jth student. The sum of what she gets from muffin 1, muffin 2, ..., muffin m. is $\frac{m}{s}$. Hence $\sum_{i=1}^{m} x_{ij} = \frac{m}{s}$.
(5) For all $1 \le i \le m, 1 \le j \le s, z \le x_{ij}$.
(6) Maximize z.

This does not work. The problem is that (say) x_{13} could be 0. In fact, it is likely that some x_{ij} is 0. This makes $z = 0$. What we really want is

$$x_{ij} \ne 0 \implies x_{ij} \ge z.$$

In any (m, s)-procedure there will be a (possibly empty) set of pairs (i, j) such that nothing from muffin i goes to student j. Hence the following *insane* procedure will find $f(m, s)$ and at the same time prove it exists and is rational.

For all $X \subseteq \{x_{11}, \ldots, x_{ms}\}$

In MUFFINLP set all of the variables of X to 0.
Solve this LP. Call the answer $f(m, s)_X$

Output

$$\max\{f(m, s)_X : X \subseteq \{x_{11}, \ldots, x_{ms}\}\}.$$

Appendix C: f(m, s) Exists! f(m, s) is Rational! f(m, s) is Computable!

By Lemma C.2 (1) $f(m, s)_X$ exists since the LP bounds all the variables between 0 and 1 and (2) $f(m, s)_X \in \mathbb{Q}$. □

Note C.8. The algorithm in Theorem C.7 is not practical since there are many X's to try. **Do Not Use!**

C.5. A Proof That Uses Mixed Integer Programming

The following proof was independently discovered by Veit Elser and Robert Fleischman.

Theorem C.9.

(1) *For all m, s, $f(m, s)$ exists.*
(2) *For all m, s, $f(m, s) \in \mathbb{Q}$.*
(3) *The function $f(m, s)$ is computable.*

Proof. We will take the MUFFINLP and add some integer-variables and constraints.

It is easy to show that $f(m, s) \geq \frac{1}{s}$. Hence every nonzero x_{ij} is $\geq \frac{1}{s}$. We will use this in our proof.

Take MUFFINLP and modify it to form MUFFINMIP.

For $1 \leq i \leq m, 1 \leq j \leq s$,

(1) Add integer variable y_{ij} and constraints $y_{ij} \leq 1$ and $y_{ij} \geq 0$ so that $y_{ij} \in \{0, 1\}$.
(2) Add the constraint $x_{ij} + y_{ij} \leq 1$.
(3) Add the constraint $x_{ij} + y_{ij} \geq \frac{1}{s}$.
(4) Replace the constraint $z \leq x_{ij}$ with $z \leq x_{ij} + y_{ij}$.
(5) We still want to maximize z.

Assume that we have a solution x_{ij}, y_{ij}. We show that this represents an optimal (m, s)-procedure. We need that z is $\leq$ every *nonzero* x_{ij} and *not* have it $\leq$ any of the variables that are set to 0. There are cases:

Case 0: $x_{ij} = 0$. We look at what the constraints tell us.

- $x_{ij} + y_{ij} \leq 1$. Since $x_{ij} = 0$ and $y_{ij} \in \{0, 1\}$, this imposes no constraint on y_{ij}.

- $x_{ij} + y_{ij} \geq \frac{1}{s}$. Since $x_{ij} = 0$ and $y_{ij} \in \{0, 1\}$, $y_{ij} = 1$.
- $z \leq x_{ij} + y_{ij}$ is satisfied since $y_{ij} = 1$. Hence x_{ij} has no effect on z.

Case 1: $x_{ij} > 0$. We look at what the constraints tell us.

- $x_{ij} + y_{ij} \leq 1$. Since $x_{ij} > 0$ and $y_{ij} \in \{0, 1\}$, this implies $y_{ij} = 0$.
- $x_{ij} + y_{ij} \geq \frac{1}{s}$. Since $y_{ij} = 0$, this implies $x_{ij} \geq \frac{1}{s}$. This makes sense since the muffin problem always has the trivial solution of dividing all muffins into s pieces of size $\frac{1}{s}$ and giving everyone m of those pieces.
- $z \leq x_{ij} + y_{ij}$ means $z \leq x_{ij}$ since $y_{ij} = 0$.

By Lemma C.5, the resulting MIP can be solved and has a rational solution. □

Note C.10. The MIP program described in Theorem C.9 is very slow. In this book, we gave several ways to find upper bounds for $f(m, s)$. If you add an upper bound on z to the MIP then it will run faster. We have done this and been able to solve some moderately large muffin problems. Using a free MIP package, we could get up to around 50 muffins, 19 students.

C.6. A Proof That Uses Topology

The following proof was discovered by Caleb Stanford in 2013 (who also helped me with this writeup). The proof is very different from those in Theorems C.7 and C.9 since it uses elementary topology rather than LPs or MIPs. The proof shows that $f(m, s)$ exists and is rational; however, it does not show that $f(m, s)$ is computable.

Definition C.11. Let $p_1, p_2, \ldots$ be a sequence of points in $\mathbb{R}^n$. Let $A \subseteq \mathbb{R}^n$.

(1) $p_1, p_2, \ldots$ is *bounded* if there exist positive numbers $B_1, \ldots, B_n$ such that, for $1 \leq i \leq n$, for $j \in \mathbb{N}$, the ith component of p_j is between $-B_i$ and B_i.

(2) A *limit point* of $p_1, p_2, \ldots$ is a point p such that, for all ϵ, for all i, there is a $j > i$ such that p_j is within ϵ of p.

(3) Let $A \subseteq \mathbb{R}^n$. A is *closed* if the limit points of every bounded sequence of points in A are in A.

Appendix C: f(m, s) Exists! f(m, s) is Rational! f(m, s) is Computable!   201

Example C.12.

(1) The sequence $0, 1, 2, 3, \ldots$ is not bounded.
(2) The sequence $1, \frac{1}{2}, \frac{1}{3}, \ldots$ is bounded. It has one limit point: 0.
(3) The sequence $1, -1, 1, -1, \ldots$ is bounded. It has two limit points: 1 and -1. While they are formally limit points, this is not what the definition was intended to capture. Limit points are intended to capture the notion of getting closer and closer to a point, not of hitting the point exactly infinitely often.
(4) The sequence $1 + \frac{1}{2}, 10 + \frac{1}{2}, 1 + \frac{1}{3}, 10 + \frac{1}{3}, \ldots$ is bounded. It has two limit points: 1 and 10.
(5) The set $(0, 2)$ is not closed. The sequence $1, \frac{1}{2}, \frac{1}{3}, \ldots$ is a bounded sequence of points in the set, whose limit point, 0, is not in the set.
(6) The set $[0, 1]$ is closed. We do not prove this.
(7) The sequence $(0, 0), (\frac{1}{2}, -\frac{1}{2}), (\frac{3}{4}, -\frac{3}{4}), \ldots,$ has one limit point at $(1, -1)$.

The following is well known.

Lemma C.13. *Let $m, n \in \mathbb{N}$. Let M be an $n \times m$ matrix and $\vec{b} \in \mathbb{R}^m$.*

(1) $[0, 1]^n$ *is a closed and bounded set.*
(2) $\{\vec{x} : M\vec{x} \leq \vec{b}\}$ *is a closed set.*
(3) *The intersection of a closed and bounded set with a closed set is closed and bounded.*
(4) $[0, 1]^n \cap \{\vec{x} : M\vec{x} \leq \vec{b}\}$ *is closed and bounded. (This follows from items (1), (2), and (3) above.)*
(5) *A finite union of disjoint closed and bounded sets is closed and bounded.*
(6) *Let X be a closed and bounded set. Let F be a continuous function from X to $\mathbb{R}$. Then there is a point $x \in X$ such that $F(x)$ is the max value of $\{F(y) : y \in X\}$.*

Theorem C.14.

(1) *For all m, s, $f(m, s)$ exists.*
(2) *For all m, s, $f(m, s) \in \mathbb{Q}$.*

Proof. (1) We prove $f(m, s)$ exists. Note that you cannot extract from this proof an algorithm to find $f(m, s)$.

Following the proof of Theorem C.7, and using that in an optimal solution all pieces are $\geq \frac{1}{s}$, we note that any (m, s)-procedure of interest can be viewed as a way to set the variables x_{ij} as $1 \leq i \leq m$, $1 \leq j \leq s$ such that:

(1) $\frac{1}{s} \leq x_{ij} \leq 1$ of $x_{ij} = 0$.
(2) For each $1 \leq i \leq m$, $\sum_{j=1}^{s} x_{ij} = 1$.
(3) For each $1 \leq j \leq s$, $\sum_{i=1}^{m} x_{ij} = \frac{m}{s}$.

Let $\vec{x}$ be the vector of all variables x_{ij}. Let M be the matrix, and $\vec{b}$ be the vector, such that constraints (2) and (3) can be expressed as $M\vec{x} \leq \vec{b}$. We define the following set:

$$\text{POSS} = \left(\left[\frac{1}{s}, 1 \right] \cup \{0\} \right)^{ms} \cap \{\vec{x} : M\vec{x} \leq \vec{b}\}.$$

By Lemma C.13, POSS is a closed set. POSS is a finite set of disconnected components. We give an example of one such component:

$\text{POSS}((1, 3), (2, 2)) = \text{POSS} \cap$

$$\left\{ \vec{x} : x_{13} = 0 \land x_{22} = 0 \land \text{ for all } (i, j) \neq (1, 3), (2, 2) \ x_{ij} \geq \frac{1}{s} \right\}.$$

For every subset of $\{(i, j) : 1 \leq i \leq m, 1 \leq j \leq s\}$, there is a component that sets those indexed variables to 0 and the rest are required to be $\geq \frac{1}{s}$. Hence there are 2^{ms} connected components. By Lemma C.13, each of the components is a closed and bounded set.

POSS is the set of all possible ways to divide up m muffins and give them to s students so that every student gets $\frac{m}{s}$ and each piece is $\geq \frac{1}{s}$. The vectors in POSS with the maximum minimal entry represent the optimal procedures.

Let $F : \text{POSS} \to \mathbb{R}^+$ map every vector in POSS to its minimum nonzero element. It is easy to see that F is continuous on POSS. By Theorem C.14, there exists an $\vec{x}$ such that $F(\vec{x})$ is the max value of $\{F(\vec{y}) : \vec{y} \in \text{POSS}\}$. This $\vec{x}$ is an optimal way to divide the muffins, and $F(\vec{x}) = f(m, s)$. Hence $f(m, s)$ exists.

(2) We proof that $f(m, s) \in \mathbb{Q}$. We *could* try a proof where we show that $f(m, s) \notin \mathbb{Q}$ leads to a contradiction. We are not going to do that directly. We are going to show that if the optimal $\vec{x}$ has some irrational elements

then we can find another optimal $\vec{x}'$ that has fewer irrational elements. By repeating this procedure, we obtain an optimal $\vec{x}$ with all rational elements.

We view the values of x_{ij} as a matrix A:

$$\begin{pmatrix} x_{11} & \cdots & x_{1s} \\ \vdots & \ddots & \vdots \\ x_{m1} & \cdots & x_{ms} \end{pmatrix}.$$

Note that each row of the matrix sums to $1 \in \mathbb{Q}$ and each column sums to $\frac{m}{s} \in \mathbb{Q}$. We first show the following which will not yield our theorem; however, it will be instructive.

Claim 1: If $f(m, s) \in \mathbb{Q}$, then there is an optimal matrix A with all rational elements.

Proof of Claim 1:
Let A be an optimal solution with some irrational entries. We show that there is an optimal solution with less irrational entries. The claim follows by iterating the process. Note that this *is not* a proof by contradiction.

We state the following obvious fact since it will have an analog in Claim 2:

- If a row has an element in $\mathbb{R} - \mathbb{Q}$, then it must have another element in $\mathbb{R} - \mathbb{Q}$: if not, then the sum of the row would not be rational.
- If a column has an element in $\mathbb{R} - \mathbb{Q}$, then it must have another element in $\mathbb{R} - \mathbb{Q}$: if not, then the sum of the column would not be rational.

Since A is an optimal solution with some irrational entries there exists $1 \le i_1 \le m$, $1 \le j_1 \le s$, $\beta_1 \notin \mathbb{Q}$ such that $\beta_1 = x_{i_1, j_1}$. Since $\beta_1 \notin \mathbb{Q}$ there must be another irrational element β_2 in row i_1. Let $\beta_2 = x_{i_1, j_2}$. Form a sequence of triples: (i_k, j_k, β_k) such that:

(1) For all k, $\beta_k = x_{i_k, j_k}$.
(2) For all k, $\beta_k \notin \mathbb{Q}$.
(3) If k is odd then $i_{k+1} = i_k$.
(4) If k is even then $j_{k+1} = j_k$.

Since A is finite, the sequence eventually repeats. Replace the sequence with only the finite repeating portion (which will be an even-length sequence). Renumber so that β_1 is the minimum irrational in the sequence. Let

$$\beta_1 - f(m, s) = \epsilon.$$

Since $f(m, s) \in \mathbb{Q}$, $\epsilon > 0$. (This is where we use $f(m, s) \in \mathbb{Q}$. Proving the analogous theorem with $f(m, s) \notin \mathbb{Q}$ is our main goal.) Let $0 < \delta < \epsilon$ be such that $\beta_1 - \delta \in \mathbb{Q}$. If we subtract δ from all of the odd-indexed β's, and add δ to all the even-indexed β's, then we have another solution with minimum element $f(m, s)$ and at least one less irrational.

End of Proof of Claim 1

The proof above will not work if $f(m, s) \notin \mathbb{Q}$. The problem is that in this case, β_1 may be equal to $f(m, s)$, so we can't necessarily choose $\delta > 0$ and subtract it from each odd-indexed β while preserving that each nonzero element of the matrix is at least $f(m, s)$. To proceed, we should require a further restriction when we find the sequence β_i: instead of just requiring these values to be irrational, we also want to exclude every odd-indexed element from being equal to $f(m, s)$. We do this by defining sets q1POS and q1NEG (you will see the reason for the names soon) such that $f(m, s) \in$ q1POS $-$ q1NEG. We will find a sequence such that the odd-indexed β's are in q1POS and the even-indexed βs are in q1NEG.

We now proceed with the proof.

Claim 2: If $f(m, s) \notin \mathbb{Q}$, then there is an optimal matrix A with all rational elements (a contradiction).

Proof of Claim 2:
We show that if there is an optimal solution with irrational entries then there is an optimal solution with less irrational entries. The claim follows by iterating the process. Note that this *is* a proof by contradiction since if every entry is in $\mathbb{Q}$ then $f(m, s)$ is not in the matrix.

Let $\alpha_1, \alpha_2, \ldots, \alpha_N, \alpha_{N+1}$ be such that:

(1) $\alpha_1, \ldots, \alpha_N, \alpha_{N+1}$ are linearly independent over $\mathbb{Q}$.
(2) All elements of A are linear combinations, with coefficients in $\mathbb{Q}$, of
    $\alpha_1, \ldots, \alpha_N, \alpha_{N+1}$.
(3) $\alpha_1 = f(m, s)$ and $\alpha_{N+1} = 1$.

Appendix C: f(m, s) Exists! f(m, s) is Rational! f(m, s) is Computable!

Let

$$\text{q1POS} = \left\{ q_0 + \sum_{i=1}^{N} q_i \alpha_i : q_0, \ldots, q_N \in \mathbb{Q}, q_1 > 0 \right\},$$

$$\text{q1NEG} = \left\{ q_0 + \sum_{i=1}^{N} q_i \alpha_i : q_0, \ldots, q_N \in \mathbb{Q}, q_1 < 0 \right\}.$$

Note the following:

- q1POS ∩ q1NEG = ∅: This follows from the α_i's being linearly independent over $\mathbb{Q}$.
- $f(m, s) \in$ q1POS: Take $q_0 = q_2 = q_3 = \cdots = q_N = 0$ and $q_1 = 1$.
- $f(m, s) \notin$ q1NEG: This follows from q1POS ∩ q1NEG = ∅ and $f(m, s) \in$ q1POS.
- If a row has an element of q1POS, then it has to have an element of q1NEG: If not, then the row won't sum to a rational.
- If a column has an element of q1POS, then it has to have an element of q1NEG: If not, then the column won't sum to a rational.

Since A is an optimal solution and $f(m, s) \in$ q1POS, there is $1 \le i_1 \le m$, $1 \le j_1 \le s$, $\beta_1 \in$ q1POS, such that $\beta_1 = x_{i_1, j_1}$. Since $\beta_1 \in$ q1POS, there must be an element of q1NEG in row i_1, say $\beta_2 = x_{i_1, j_2}$. Form a sequence of triples: (i_k, j_k, β_k) such that:

(1) For all k, $\beta_k = x_{i_k, j_k}$.
(2) If k is odd then $\beta_k \in$ q1POS.
(3) If k is even then $\beta_k \in$ q1NEG.
(4) If k is odd then $i_{k+1} = i_k$.
(5) If k is even then $j_{k+1} = j_k$.

Since A is finite, the sequence eventually repeats. Replace the sequence with only the finite repeating portion (which will be an even-length sequence). Renumber so that β_2 is the minimum even-indexed value in the sequence, so $\beta_2 \in$ q1NEG. Let

$$\beta_2 - f(m, s) = \epsilon.$$

Since $f(m,s) \in$ q1POS and $\beta_2 \in$ q1NEG, $\beta_2 \neq f(m,s)$ so $\epsilon > 0$. (This last line is the key to the proof and the reason we defined q1POS and q1NEG.) Let $0 < \delta < \epsilon$ be such that $\beta_2 - \delta \in \mathbb{Q}$. If we subtract δ from all of the even-indexed β's and add δ to all the odd-indexed β's then we have another solution with minimum element at least $f(m,s)$ and at least one less irrational.

End of Proof of Claim 2 □

References

Aziz, H. and Mackenzie, S. (2016). A discrete and bounded envy-free cake cutting protocol for any number of agents, in *FOCS16* (IEEE Computer Society Press), pp. 416–427, https://arxiv.org/abs/1604.03655.

Brams, S. J. and Taylor, A. D. (1995). An envy-free cake division protocol, *The American Mathematical Monthly* **102**, pp. 9–18.

Brams, S. J. and Taylor, A. D. (1996). *Fair Division: From Cake-cutting Do Dispute Resolution* (Cambridge University Press).

Chatwin, R. (2019). An optimal solution for the muffin problem, https://arxiv.org/abs/1907.08726.

Cui, G., Dickerson, J., Durvasula, N., Gasarch, W., Metz, E., Prinz, J., Raman, N., Smolyak, D., and Yoo, S. H. (2018). A muffin-theorem generator, in *Ninth International Conference on Fun with Algorithms*, pp. 15.1–15.19, http://drops.dagstuhl.de/opus/portals/lipics/index.php?semnr=16069.

Edmonds, J. and Pruhs, K. (2011). Cake cutting really is not a piece of cake, *ACM Transactions on Algorithms* Previously appeared in *SODA 2006*.

Even, S. and Paz, A. (1984). A note on cake cutting, *Discrete Applied Mathematics* **7**, pp. 285–296.

Gasarch, W., Huddleston, S., Metz, E., and Prinz, J. (2019). Complexity Theory Column 100: The Muffin Problem, *SIGACT News* **50**, 2, pp. 31–60, this is Lane Hemaspaandra's complexity column.

Robertson, J. and Webb, W. (1998). *Cake Cutting Algorithms: Be Fair If You Can* (A. K. Peters).

Index

$\lceil x \rceil$, 7
$\lfloor x \rfloor$, 7
$\{L\ a\}$, 22
($\exists$), 190
($\forall$), 190
(a, b), 182
$(a, b]$, 182
(m, s)-procedure, 11
$A \subseteq B$, 183
$A \supseteq B$, 183
$B(x)$, 8
V-share, 45
V-student, 45
V-conjecture, 60, 65
VV-conjecture, 81
$[a, b)$, 182
$[a, b]$, 182
$\mathbb{C}$, 182
$FC(m, s)$, 21
$Half(m, s)$, 64
$\mathbb{N}$-solution, 29
$\mathbb{N}$, 182
$\mathbb{Q}$, 182
$\mathbb{R}$, 182
$VHalf(m, s, \alpha)$, 63
$\mathbb{Z}$, 182
$\cap$, 182
$\cup$, 182
$\emptyset$, 182
$\in$, 181
$\notin$, 181

$\sum$, 191
$f(m, 1)$, 14
$f(m, 2)$, 14
$f(m, 3)$, 23
$f(m, 4)$, 24
$f(m, 5)$, 69
$f(m, s)$, 11
$f(m, s) \geq \alpha$, 12
$f(m, s) \leq \alpha$, 12

A

Alice, 3

B

Bachman, Nancy, v
bijection, 185
Bob, 3
buddy, 8
buddy–match sequence, 99
buddying, 98

C

ceiling, 7
Chatwin, Richard, vii, 22
Choi, Yunseo, 6, 105
closed interval, 182
computable, 11
Cong, Kevin, 6, 105
convention about intervals, 87
Copeland, Jeremy, v

D
Duality Theorem, 15

E
Elser, Veit, viii, 199

F
Fair Division, viii
Fair Division, envy-free, 194
Fair Division, proportional, 194
Fair Divisions, discrete algorithms, 194
floor, 7
Floor–Ceiling Theorem, 19–20
for all (($\forall$), 190
formula for $f(m, 4)$, 24
Frank, Alan, v, viii–ix, 26
Friedman, Erich, viii, 15
FUN with Algorithms, ix

G
Gardner, Martin, v
Gathering for Gardner, v

H
Huddleston, Scott, vii, 60

I
injection, 185
intersection ($\cap$), 182
interval, 182

J
Julia Robinson Mathematics Festival, v

K
Kronzek, Rochelle, v
Kruskal, Clyde, v

L
linear programming, 196

M
matching, 98
measure of our progress, 17
mixed integer programming, 197
mod, 186
Muffin Math Song, x
Muffin website, vii, 22
multiset, 184

N
notation for a student having many shares, 22
NP-complete, 197

O
open interval, 182

P
piece, 3
Pigeon Hole Principle, 6
Problems with a Point, v
Propp, James, viii

R
recreational math, vi

S
Scientific American, v
serious math, vi
sets, 181
share, 3
shares of size $\frac{1}{2}$, 87
Simplex Method, 197
Stanford, Caleb, viii
star to indicate a tight bound, 88
subset ($A \subseteq B$), 183
sum of a set, 28
superset ($A \supseteq B$), 183
surjection, 185

T
there exists (($\exists$), 190

U
union ($\cup$), 182